NOISE IN MEASUREMENTS

NOISE IN MEASUREMENTS

ALDERT VAN DER ZIEL

Electrical Engineering Department
University of Minnesota

Electrical Engineering Department
University of Florida

A WILEY-INTERSCIENCE PUBLICATION

JOHN WILEY & SONS, New York ● London ● Toronto ● Sydney

Copyright © 1976 by John Wiley & Sons, Inc.

All rights reserved. Published simultaneously in Canada.

No part of this book may be reproduced by any means, nor transmitted, nor translated into a machine language without the written permission of the publisher.

Library of Congress Cataloging in Publication Data:

Van der Ziel, Aldert, 1910–
  Noise in measurements.

  "A Wiley-Interscience publications."
  Includes bibliographical references.
  1. Electric measurements. 2. Electronic measurements. 3. Random noise theory. 4. Electric noise. 5. Electronic noise. I. Title.
QC535.V35   621.37   76-12108
ISBN 0-471-89895-3

Printed in the United States of America

10 9 8 7 6 5 4 3 2 1

PREFACE

This book discusses the effect of noise on the accuracy of measurements. It brings together material that was spread throughout many textbooks and a vast amount of past and current literature. The text was developed from a series of lectures given at the Universities of Minnesota and Florida.

Chapters 1–7 provide the necessary background. After a short introduction, the method of distribution functions for calculating averages, autocorrelation, and cross-correlation functions is developed in Chapter 2. Chapter 3 considers a few simple applications, and Chapter 4 examines binomial, Poisson, and normal distribution functions and develops the variance theorem. Chapter 5 discusses Fourier analysis methods and shows how spectral intensities can be calculated. Chapter 6 takes up noise characterization in two-terminal and four-terminal devices, and Chapter 7 examines flicker noise and generation-recombination noise.

Chapters 8–17 deal with applications. Chapter 8 treats measurements of small currents, voltages, and charges; Chapter 9 studies thermal radiation detectors like thermocouples and bolometers. Chapter 10 investigates photodetectors of the photoemissive, photodiode, and the classical detector types. Chapter 11 deals with photoconductive detectors, and Chapter 12 considers pyroelectric detectors and capacitive bolometers, then Chapter 13 examines noise in television pick-up tubes. Chapter 14 investigates photomixing, after which Chapter 15 deals with light amplification with electroluminescence. Chapter 16 gives a discussion of Josephson junction devices. Chapter 17 briefly examines high-energy quantum and particle detectors. The appendix derives a few formulas of the theory of ferroelectrics used in Chapter 12.

I am indebted to my graduate students at the Universities of Minnesota and Florida who helped shape the manuscript and to Mrs. van der Ziel who helped prepare it.

A. VAN DER ZIEL

Minneapolis, Minnesota
April 1976

CONTENTS

1. **Introduction**   1

2. **Distribution Functions, Averages, Autocorrelation, and Crosscorrelation Functions**   3
    1. Distribution Functions and Averages, 3
    2. Autocorrelation and Crosscorrelation Functions, 7

3. **Simple Applications**   9
    1. Noise in Electrical Measurements, 9
    2. Measurements by Counting Techniques, 13

4. **Typical Distribution Functions: the Variance Theorem**   17
    1. Typical Distribution Functions, 17
    2. The Variance Theorem, 22

5. **Fourier Analysis of Fluctuating Quantities: Spectral Intensities**   30
    1. Fourier Analysis; Wiener–Khintchine Theorem, 30
    2. Evaluation of $S_x(f)$ or $S_x(0)$, 37

6. **Noise Characterization Devices and Amplifiers**   50
    1. Noise Characterization, 50
    2. Application to FET Circuits, 60
    3. Noise in Transistor Circuits, 69

7. **Flicker Noise and Generation–Recombination Noise**   77
    1. Derivation of Flicker Noise Formulas from Generation–Recombination Noise, 77

2. Flicker Noise in MOSFETs, 80
3. Flicker Noise in Transistors, 82
4. Flicker Noise in Carbon Resistors, 84
5. Generation–Recombination Noise in JFETs, 85

8. Measurements of Small Currents, Voltages, and Charges — 88

1. Current Measurements, 88
2. Direct-Current Voltage Measurements, 90
3. Measurement of Small Charges, 94

9. Thermal Radiation Detectors — 99

1. General Considerations, 99
2. The Thermocouple Detector, 106
3. The Resistive Bolometer, 111

10. Photoelectric Detectors and Classical Detectors — 119

1. Photoemissive Diodes, Photodiodes, and Photovoltaic Cells, 119
2. Bandwidth Considerations, 129
3. Multiplication Methods in Junction Diodes, 133
4. Multiplication in Photoemissive Devices, 138
5. Classical Detectors, 140

11. Photoconductive Detectors — 146

1. Photoconductive Response, 146
2. Examples, 152
3. Thermal Noise, Flicker Noise, and Amplifier Noise, 155
4. Noise-Reduction Methods, 157
5. Practical Examples of Photoconductors, 159

12. Pyroelectric Detectors and Capacitive Bolometers — 160

1. The Pyroelectric Detector, 160
2. The Capacitive Bolometer, 166
3. The Nature of the Device Noise, 172

13. Noise in Television Pickup Tubes — 174

1. The Image Orthicon, 174
2. The Vidicon, 177
3. The Secondary Electron Conduction Vidicon, 179

    4. Solid-State Image Sensors, 181
    5. The Pyroelectric Vidicon, 186

14. Photomixing   188

    1. Noise-Equivalent Power of a Heterodyne Receiver, 188
    2. Photomixing, 189
    3. The Point-Contact Schottky Barrier Diode Mixer, 194

15. Light Amplification with Cathodeluminescence   198

    1. Noise in Cathodeluminescent Light, 198
    2. Noise in Light Amplifiers, 202

16. Josephson Junction Devices   204

    1. The Josephson Junction as a Thermometer, 205
    2. The Josephson Junction as an Amplifier, 209
    3. The SQUID, 213

17. High-Energy Quantum and Charged-Particle Detectors   217

    1. Principles, 217
    2. Applications, 221

Appendix. Introduction to the Theory of Ferroelectrics   223

Index   225

NOISE IN MEASUREMENTS

1

INTRODUCTION

In physics and electrical engineering one often encounters fluctuating signals generated in electrical circuits, electrical devices, or other measuring systems. Such fluctuating signals are generally called *noise*.

The name "noise" requires explanation. If the fluctuating voltage or current generated in a circuit component or electronic device is amplified by a low-frequency amplifier and the amplified signal is fed into a loudspeaker, the loudspeaker produces a hissing sound; hence the name "noise." The name "noise" now refers to any spontaneous fluctuation, independent of whether an audible sound is produced.

Noise sets a lower limit to the signals that can be processed electronically. In the same way it sets lower limits to practically all types of measurement. It is important to minimize the noise-to-signal ratio in any such measurement and so determine the limit of accuracy of these measurements. It is the aim of this book to familiarize the reader with these problems in the measurement of currents, voltages, charges, and amounts of radiation.

The important sources of noise that will be encountered are thermal noise, shot noise, generation–recombination noise, temperature-fluctuation noise, and flicker noise. We now discuss these noise sources in somewhat greater detail.

Thermal noise is due to the random motion of carriers in any conductor; as a consequence of this random motion a fluctuating electromotive force (e.m.f.) $V(t)$ is developed across the terminals of the conductor. The same phenomenon occurs in the conducting channel of field-effect transistors (FET). It is the dominant noise source in any device that is electrical in nature and in thermal equilibrium with a temperature bath kept at a fixed temperature T.

Shot noise occurs whenever a noise phenomenon can be considered as a series of independent events occurring at random. For example, in the case of emission of electrons by a thermionic cathode or by a photocathode, the emission of electrons consists of a series of independent random events; hence the emission currents show shot noise. In p–n junctions and transistors the crossing of a junction by electrical carriers (electrons or holes) constitutes a series of independent random events, hence the currents in such devices show shot noise. It equally holds when transitions occur between two energy levels, such as in the generation and recombination of carriers in a semiconductor, or when photons are emitted by a laser. In each case one must ask what entities make up the series of independent random events that produce shot noise.

Generation–recombination noise occurs whenever free carriers are generated or recombine in a semiconductor material. The fluctuating rates of generation and recombination can be considered as a series of independent events occurring at random, and hence the process can be considered as a shot-noise process. However, it is also useful to consider the fluctuation δn in the carrier density n as giving rise to a fluctuation δR in the resistance R of the device. This resistance fluctuation δR can be detected by passing a d.c. current I through the sample; the current I develops a fluctuating e.m.f. $V(t) = I \delta R(t)$ across its terminals, and this e.m.f. can be amplified and measured by standard techniques.

Temperature-fluctuation noise of a small body occurs because of the fluctuating heat exchange between the body and its environment due to fluctuations in the emitted and received radiation and to fluctuations in the heat conduction. The first can be described by fluctuations in the rate of emission and absorption of quanta by the small body. The fluctuations in the heat conduction are always present, since the small body must always have some heat-conducting path (wires, connections, etc.) to its environment. When air is blown over the small body or liquid is flowing past the small body, there is also a fluctuating *heat convection*; it is not essential, however, since it can be eliminated by proper techniques.

Flicker noise can be due to various causes and is characterized by its spectral intensity (see Chapter 5). Most noise sources have a spectral intensity that is constant at low frequencies and decreases more or less rapidly above a certain "turnover" frequency that is characteristic for the noise source in question. The various forms of flicker noise have in common the condition that their spectral intensity is of the form const/f^α with α close to unity, so that their effect is most pronounced at low frequencies.

Fluctuating quantities like currents, voltages, temperatures, or numbers of carriers are called *random variables*. One speaks of a *continuous* random variable when the fluctuating quantity can assume a continuous range of values and of a *discrete* variable when the fluctuating quantity can only assume discrete values. The fluctuating number of carriers in a semiconductor sample is a discrete random variable.

2

DISTRIBUTION FUNCTIONS, AVERAGES, AUTOCORRELATION, AND CROSSCORRELATION FUNCTIONS

In calculations about noise in electrical measuring systems one must often calculate the averages of a function $g(X)$ of the random variable $X(t)$ in question. It is denoted by $\overline{g(X)}$ and is calculated with the help of the *probability density function* or distribution function of the variable $X(t)$.

This function, in turn, is introduced by considering probabilities in an *ensemble*, namely a very large assembly of systems subjected to independent fluctuations. To make the discussion more precise, the number of systems should go to infinity.* We discuss this in Section 2.1a for a single random variable and in Section 2.1b for multiple random variables.

In the case of two random variables $X(t)$ and $Y(t)$ with $\overline{X} = \overline{Y} = 0$, the average $\overline{XY}$ may not be zero. The quantities $X(t)$ and $Y(t)$ are then said to be *correlated*. A particular case of correlation occurs if we consider a random variable $X(u)$ at the instants t and $(t+s)$, the function $\overline{X(t)X(t+s)}$ is called the autocorrelation *function*. Extension to several random variables leads to *autocorrelation* and *crosscorrelation functions*. (Section 2.2).

2.1 DISTRIBUTION FUNCTIONS AND AVERAGES

2.1a Single Random Variable

We consider an ensemble of N systems in which the fluctuations are described by the random variable $X(t)$ and let N go to infinity. Let ΔN elements of the ensemble have a value of $X(t)$ between X and $(X + \Delta X)$ at

---

*In an ensemble with N elements the relative accuracy of the averages is $N^{-1/2}$, so that it corresponds to 0.01 for $N = 10^4$.

the instant t_1. One then calls $\Delta P = (\Delta N/N)$ the probability that the random variable $X(t)$ had a value between X and $(X+\Delta X)$ at the instant t_1. Obviously ΔN is proportional to ΔX as long as ΔX is sufficiently small, so that $(\Delta P/\Delta X)$ is independent of ΔX. More precisely, we may write in differential form

$$\frac{dP}{dX} = f(X,t_1), \quad \text{or} \quad dP = f(X,t_1)dX \qquad (2.1)$$

The function $f(X,t_1)$ is called the *probability density function* of X at the instant t_1. When $f(X,t_1+t)$ is independent of t, that is,

$$f(X,t_1+t) = f(X,t_1) = f(X) \qquad (2.1a)$$

the variable is said to be *stationary*. The noise processes encountered in physics and engineering are nearly always stationary.

Since the variable X must certainly lie within the range of allowed values, we have, if the integration is extended over all allowed values of X,

$$\int f(X)dX = 1 \qquad (2.2)$$

Such a function $f(X)$ is said to *normalized*. If $f(X)$ is not normalized, it can be multiplied by a normalizing factor C so that $Cf(X)$ is normalized, that is,

$$\int Cf(X)dX = 1 \quad \text{or} \quad C = \left[\int f(X)dX\right]^{-1} \qquad (2.2a)$$

We may thus assume without lack of generality that $f(X)$ is normalized.

We can now define *ensemble averages* as follows: The ensemble average of X^m, denoted by $\overline{X^m}$, is defined as

$$\overline{X^m} = \int X^m f(X)dX \qquad (2.3)$$

and the average of a function $g(X)$ of X is defined as

$$\overline{g(X)} = \int g(X)f(X)dX \qquad (2.3a)$$

where the integration is extended over all values of X. If $f(X)$ is symmetrical in X, that is, if $f(X) = f(-X)$, and X can vary between $-X_0$ and X_0, then the averages of all odd powers of X are zero.

The most important averages are $\overline{X}$ and $\overline{X^2}$. If $\overline{X}$ is not zero, one should introduce $\Delta X = (X - \overline{X})$ as a new random variable. The most important

Distribution Functions and Averages

average is then $\overline{\Delta X^2}$, which is denoted by the symbols $\operatorname{var} X$ or σ_x^2,

$$\operatorname{var} X = \sigma_x^2 = \overline{(X-\overline{X})^2} = \overline{X^2} - \overline{2X\overline{X}} + (\overline{X})^2 = \overline{X^2} - (\overline{X})^2 \quad (2.4)$$

If we look at a single element of the ensemble for the time interval $0 \leqslant t \leqslant T$, we can make up the time average $\langle g(X) \rangle$ of a function $g(X)$ of X by the definition

$$\langle g(X) \rangle = \lim_{T \to \infty} \frac{1}{T} \int_0^T g(X) \, dt \quad (2.5)$$

If this time average approaches the ensemble average (2.3a) in the limit when T goes to infinity, the noise processes under investigation are said to be *ergodic*. The noise processes encountered in physics and engineering are practically always ergodic.

In the case of discrete random variables the definitions must be properly modified and all integrations must be replaced by summations. Let $P(n)$ be the probability that the discrete variable has the value n, then the normalization condition becomes

$$\sum_n P(n) = 1 \quad (2.6)$$

and the ensemble average of n^m must be defined as

$$\overline{n^m} = \sum_n n^m P(n) \quad (2.7)$$

($m = 1, 2 \ldots$). The variance of n is again defined as

$$\operatorname{var} n = \overline{(n - \overline{n})^2} = \overline{n^2} - (\overline{n})^2 \quad (2.8)$$

2.1b Multivariate Distributions and Averages

For two continuous variables $X_1(t)$ and $X_2(t)$ one can evaluate the probability that $X_1(t)$ has a value between X_1 and $(X_1 + dX_1)$ and that simultaneously $X_2(t)$ has a value between X_2 and $(X_2 + dX_2)$ at the instant t_1. In analogy with (2.1) the *joint probability* dP may be then written

$$dP = f(X_1, X_2, t_1) \, dX_1 \, dX_2 \quad (2.9)$$

and $f(X_1, X_2, t_1)$ is called the *joint probability density function* for the variables X_1 and X_2 at the instant t_1. Usually

$$f(X_1,X_2,t_1+t)=f(X_1,X_2,t_1)=f(X_1,X_2) \tag{2.9a}$$

for all values of t; the noise process is then said to be *stationary*.

The normalization condition is now

$$\int\int f(X_1,X_2)dX_1dX_2=1 \tag{2.10}$$

and averages are defined in the same manner as for single variables, that is,

$$\overline{X_1^n X_2^m} = \int\int X_1^n X_2^m f(X_1,X_2)dX_1dX_2 \tag{2.11}$$

where the integration extends over all values of X_1 and X_2.

Usually $\overline{X_1}=\overline{X_2}=0$; the most important averages are then $\overline{X_1^2}$, $\overline{X_2^2}$, and $\overline{X_1X_2}$. If $\overline{X_1X_2}=0$, the quantities are said to be *uncorrelated*; if $\overline{X_1X_2}\neq 0$, the quantities are said to correlated; the parameter

$$c = \frac{\overline{X_1X_2}}{\left(\overline{X_1^2}\cdot\overline{X_2^2}\right)^{1/2}} \tag{2.12}$$

is called the *correlation coefficient*. Applying the fact that $\overline{(aX_1+bX_2)^2}\geq 0$ for all values of a and b, it can be shown that $-1\leq c\leq 1$. The case $|c|=1$ is called *full* correlation; the case $|c|<1$ is called *partial* correlation.

The considerations are easily extended to m variables $X_1,X_2...X_m$ or to discrete variables $n_1,n_2...n_m$. The definitions are similar to the two-variable case.

If two random variables $X(t)$ and $Y(t)$ are partly correlated, namely if $|c|<1$, then one can split Y into a part aX that is fully correlated with X and a part Z that is uncorrelated with X. That is, we may write

$$Y = aX + Z \tag{2.13}$$

where $\overline{X}=\overline{Y}=\overline{Z}=0$ and $\overline{XZ}=0$. Since $\overline{XY}=a\overline{X^2}$ and $\overline{Y^2}=(a^2\overline{X^2}+\overline{Z^2})$, we have from the definition of c

$$a = c\left[\frac{\overline{Y^2}}{\overline{X^2}}\right]^{1/2} \qquad \overline{Z^2} = \overline{Y^2}(1-c^2) \tag{2.13a}$$

These formulas are useful in the discussion of noise in bipolar transistors and FETs.

2.2 AUTOCORRELATION AND CROSSCORRELATION FUNCTIONS

A particularly useful case of two partly correlated variables occurs when $X_1(t)=X(t)$ and $X_2(t)=X(t+s)$. Then the joint probability density function $f(X_1,X_2)$ can be introduced and averages can be defined in the usual manner. The average $\overline{X_1 X_2}=\overline{X(t)X(t+s)}$ is called the *autocorrelation function*; it measures how long a given fluctuation persists at later times.

The autocorrelation function has the following properties

1. $\overline{X(t)X(t+s)}$ is independent of t if $X(t)$ is stationary.
2. $\overline{X(t)X(t+s)}$ is either continuous or a δ function in s. If $\overline{X(t)X(t+s)}$ is not a δ function in s, then any discontinuities in $X(t)$ and $X(t+s)$ occur at different instants for different elements of the ensemble so that they are averaged out in the averaging process. As a consequence $\overline{X(t)X(t+s)} = \overline{X^2(t)}$ for $s=0$, unless the autocorrelation function is a δ function in s.
3. $\overline{X(t)X(t+s)}$ is symmetrical in s, if $X(t)$ is stationary. The reason is as follows:

$$\overline{X(t)X(t+s)} = \overline{X(u-s)X(u)} = \overline{X(u)X(u-s)} = \overline{X(t)X(t-s)}$$

The first step comes about by putting $u=(t+s)$. The second step is an interchange of terms. The third step involves replacing u by t, which is allowed since $X(t)$ is stationary.

4. For $s\to\infty$, $\overline{X(t)X(t+s)}$ goes to zero sufficiently fast, so that

$$\int_{-\infty}^{\infty} |\overline{X(t)X(t+s)}|\,ds \qquad (2.14)$$

exists. This is the case for all practical noise sources, except perhaps flicker noise.

5. The correlation coefficient

$$c(s) = \frac{\overline{X_1 X_2}}{\left[\overline{X_1^2} \cdot \overline{X_2^2}\right]^{1/2}} = \frac{\overline{X(t)X(t+s)}}{\overline{X^2}} \qquad (2.15)$$

is called the *normalized* autocorrelation function; it exists if $\overline{X(t)X(t+s)}$ is not a δ function in s. Here we have made use of the fact that $X(t)$ is stationary, so that $\overline{X_1^2}=\overline{X_2^2}=\overline{X^2}$.

In the particular case of two partially correlated quantities $X(t)$ and $Y(t)$ one can introduce the *autocorrelation* functions $\overline{X(t)X(t+s)}$ and $\overline{Y(t)Y(t+s)}$ and the *crosscorrelation* functions $\overline{X(t)Y(t+s)}$ and

$\overline{X(t+s)Y(t)}$. The first are symmetrical in s, whereas the latter are usually not. Moreover, the crosscorrelation functions, although related, are not identical. We have in analogy with the case of a single variable

$$\overline{X(t)Y(t+s)} = \overline{X(u-s)Y(u)} = \overline{X(t-s)Y(t)} \qquad (2.16)$$

$$\overline{X(t+s)Y(t)} = \overline{X(u)Y(u-s)} = \overline{X(t)Y(t-s)} \qquad (2.16a)$$

In each case the first step replaces $(t+s)$ by u and the second replaces u by t. The latter is allowed if the noise processes are stationary.

3

SIMPLE APPLICATIONS

With the help of the concept of averages of random variables, we can already understand the limiting accuracy of some simple measurements due to the inherent noise in the measuring system.

As a first case we consider noise in electrical circuits (Section 3.1a) and show that it becomes significant when the signal to be processed lies within the μV range. As a second case we consider noise in sensitive galvanometers (Section 3.1b) and show that it limits the accuracy of current measurements when the currents are of the order of 1 pA.

Much more sensitive measurements can be performed when counting techniques can be introduced (Section 3.2). For example, electron currents in a vacuum can be multiplied by an electron multiplier so that individual electrons can be counted; very small currents can be measured in this manner (Section 3.2a). Or a stream of photons can be made to impinge upon the photocathode of a photomultiplier and the individual photons can be counted; very small radiant powers can be determined in this manner (Section 3.2b).

3.1 NOISE IN ELECTRICAL MEASUREMENTS

When small particles are suspended in a liquid, they execute a random zigzag motion, called *Brownian motion*, named after its discoverer Robert Brown (1827). The problem was extensively studied during 1890–1910, and Einstein developed its theory in 1905. Einstein suggested that the mean kinetic energy per degree of freedom of the particle should be given by statistical mechanics as

$$\tfrac{1}{2} M \overline{v_x^2} = \tfrac{1}{2} kT \tag{3.1}$$

where M is the mass of the particle, v_x its instantaneous velocity component in the X direction, $\overline{v_x^2}$ its mean-square value, k the Boltzmann constant, and T the absolute temperature. However, what is actually *observed* under the microscope is not the instantaneous velocity v_x, but the displacement Δx in the X direction during the time interval t. Einstein showed that

$$\overline{\Delta x^2} = 2Dt \tag{3.2}$$

where $\overline{\Delta x^2}$ is the mean-square value of Δx and D is the diffusion constant of the particle. This equation was later verified experimentally.

Einstein quickly realized that many physical systems would show Brownian motion. For example, thermal noise is nothing but "Brownian motion of electricity" in electrical circuits. We discuss here the case of the input circuit of a wideband amplifier and the case of a sensitive galvanometer.

3.1a Noise in Electrical Circuits

As an example of an electrical circuit we consider the RC circuit of Fig. 3.1. As a result of the random thermal agitation of the electrons in the resistor R the capacitor C will be charged and discharged at random. In analogy with (3.1) the average energy stored in the capacitor will be

$$\tfrac{1}{2} C\overline{v^2} = \tfrac{1}{2}kT, \quad \text{or} \quad \overline{v^2} = \frac{kT}{C} \tag{3.3}$$

The value of $(\overline{v^2})^{1/2}$ is in the μV range. Take $C = 30$ pF, $T = 300°$K, $k = 1.38 \times 10^{-23}$ joule deg^{-1}, then $\overline{v^2} = (1.38 \times 10^{-10}$ V$^2)$ or $(\overline{v^2})^{1/2} \simeq 12$ μV. If we must process electronic signals with the help of wideband amplifiers, we thus run into noise problems in the input circuit if the input-signal level is in the μV range.

It should be noted that the resistor R does not enter into the final result; it only determines how wide the passband of a wideband amplifier process-

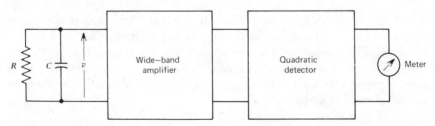

Figure 3.1. The thermal noise voltage v across an RC circuit is amplified by a wide-band amplifier, passed through a quadratic detector, and measured.

ing the noise signal must be in order to measure the full voltage developed across the capacitor C.

The measuring arrangement of $\overline{v^2}$ is shown in Fig. 3.1. The electrical circuit is connected to a wideband amplifier that amplifies the voltage v by a factor $g \gg 1$. The output voltage is fed into a quadratic detector that gives an output $g^2 \overline{v^2}$ and the output of the quadratic detector is fed into a meter M, so that the deflection of the meter is proportional to $\overline{v^2}$.

Quite often the problem is different. For example, one wants to measure a small d.c. voltage v_0 or a small single frequency a.c. signal $v_0 \cos \omega t$. The noise is then a disturbing factor and one wants to make it as small as possible. It is then important to filter most of the noise out by means of a narrow-band filter (a lowpass filter in the d.c. case and a bandpass filter of center frequency ω for the a.c. case). To evaluate the filter output one must make a Fourier analysis of the noise; that problem is discussed in Chapter 5.

We finally prove the equipartition theorem (3.3). If a system has a temperature T, the probability that it has an energy E is proportional to $\exp(-E/kT)$; the factor $\exp(-E/kT)$ is called the *Boltzmann factor*. In the RC circuit the energy stored in the capacitor C is $\frac{1}{2}Cv^2$, where v is the voltage across C. The probability dP of finding a voltage between v and $(v+dv)$ is therefore

$$dP = C_0 \exp\left(-\frac{\frac{1}{2}Cv^2}{kT}\right) dv \qquad (3.4)$$

where the constant C_0 must be so chosen that

$$\int_{-\infty}^{\infty} dP = 1 \quad \text{or} \quad C_0 \int_{-\infty}^{\infty} \exp\left(-\frac{\frac{1}{2}Cv^2}{kT}\right) dv$$

$$= C_0 \left(\frac{2kT}{C}\right)^{1/2} \int_{-\infty}^{\infty} \exp(-x^2) dx$$

$$= \left(\frac{2kT}{C}\right)^{1/2} \pi^{1/2} C_0 = 1, \quad \text{or} \quad C_0 = \left(\frac{C}{2\pi kT}\right)^{1/2} \qquad (3.4a)$$

Here $x^2 = (Cv^2/2kT)$. Therefore

$$\overline{v^2} = C_0 \int_{-\infty}^{\infty} v^2 \exp\left(-\frac{Cv^2}{2kT}\right) dv = C_0 \left(\frac{2kT}{C}\right)^{3/2} \int_{-\infty}^{\infty} x^2 \exp(-x^2) dx$$

$$= C_0 \left(\frac{2kT}{C}\right)^{3/2} \cdot \frac{\pi^{1/2}}{2} = \frac{kT}{C} \qquad (3.4b)$$

as is found by substituting for C_0.

3.1b Noise in Galvanometers

Another example of Brownian motion is the random rotation of a galvanometer coil around its axis of suspension. Let us consider a galvanometer with electromagnetic damping, where the deflection of the coil is described by the differential equation

$$\Theta \frac{d^2\varphi}{dt^2} + n_0^2 \frac{A^2 B^2}{r} \frac{d\varphi}{dt} + D\varphi = n_0 A B i(t) \qquad (3.5)$$

Here Θ is the moment of inertia of the coil, A its area, n_0 its number of turns, B the magnetic flux density, $-D\varphi$ the restoring torque that tends to bring the coil back to its equilibrium position, and $i(t)$ the current flowing through the coil, giving rise to a driving torque $n_0 A B i(t)$ acting on the coil. Finally r is the resistance of the loop circuit consisting of the coil resistance plus any external resistor.

The term $(nA^2B^2/r)(d\varphi/dt)$ represents the damping torque acting on the coil. To understand this term, we observe that $(d\varphi/dt)$ is the angular velocity of the coil; hence $-n_0 A B (d\varphi/dt)$ is the induced e.m.f. caused by the coil motion and $-(n_0 A B/r)(d\varphi/dt)$ is the current flowing through the circuit. Multiplying this expression by $n_0 A B$ to obtain the torque, we find for the damping torque $-(n_0^2 A^2 B^2/r)(d\varphi/dt)$. This torque occurs in the right-hand side of (3.5); by shifting it to the left-hand side we obtain (3.5) in its final form.

If the current $i(t)$ flowing through the coil is a direct current I, the steady-state solution of (3.5) is

$$\varphi = \frac{n_0 A B}{D} I \qquad (3.6)$$

The constant $(n_0 A B/D)$ is called the *technical sensitivity* of the galvanometer; it can be made large by making D small. However, if one does this, one runs into noise problems. For according to the equipartition theorem the average potential energy stored in the coil is

$$\tfrac{1}{2} D \overline{\varphi^2} = \tfrac{1}{2} kT, \quad \text{or} \quad \overline{\varphi^2} = \frac{kT}{D} \qquad (3.7)$$

A small value of D thus corresponds to a large root mean square (r.m.s.) deflection $(\overline{\varphi^2})^{1/2}$ of the coil. The proof of (3.7) is similar to the earlier proof of (3.3).

Unfortunately, the constant D is not so easily measured; we should therefore express $\overline{\varphi^2}$ in more accessible parameters. Moreover, we are not so much interested in $(\overline{\varphi^2})^{1/2}$ itself, but more in the minimum current I_{min} that corresponds to it.

Galvanometers are practically always used under the condition of critical damping, since this gives the best and fastest response. To explain what this means we investigate the response of the galvanometer if $i(t)$ is a step function $au(t)$, where a is a small current step. To that end we introduce three parameters: (a) the resonant frequency ω_0, (b) the damping constant n, and (c) the time constant τ_0 by the definitions

$$\omega_0 = \left(\frac{D}{\Theta}\right)^{\frac{1}{2}}; \quad \frac{n_0^2 A^2 B^2}{r} = 2n\omega_0 \Theta, \quad \tau_0 = \frac{2\pi}{\omega_0} \tag{3.8}$$

Dividing (2.5) by Θ and making the substitutions we obtain

$$\frac{d^2\varphi}{dt^2} + 2n\omega_0 \frac{d\varphi}{dt} + \omega_0^2 \varphi = \frac{n_0 A B}{\Theta} au(t) \tag{3.9}$$

A standard mathematical analysis now shows that the solution of (3.9) is damped periodic for $n < 1$ and aperiodic for $n > 1$. The fastest and best response occurs for $n = 1$; the galvanometer is then operating at the limit of aperiodicity and is said to be *critically damped*.

We now introduce a limiting d.c. current I_{min} by requiring that I_{min} gives a deflection equal to $(\overline{\varphi^2})^{1/2}$. Hence

$$I_{min} \frac{n_0 A B}{D} = (\overline{\varphi^2})^{1/2} \tag{3.10}$$

or

$$I_{min} = \left[\frac{D^2}{n_0^2 A^2 B^2} \cdot \frac{kT}{D}\right]^{1/2} = \left(\frac{DkT}{2n\omega_0 \Theta r}\right)^{1/2} = \left(\frac{\pi kT}{nr\tau_0}\right)^{1/2} \tag{3.10a}$$

These quantities are easily measured.

Taking $r = 1000 \ \Omega$, $n = 1$, $\tau_0 = 2$ s, $T = 300°$ K yields $I_{min} = 2.5 \times 10^{-12}$ A, so that galvanometer noise becomes important when measuring currents in the pA range.

3.2 MEASUREMENTS BY COUNTING TECHNIQUES

Suppose certain events, such as emission of electrons or emission of photons, occur at random instants at the average rate λ. Such events are called *Poisson* events. The question is now how to measure λ and how to determine the accuracy of the measurement.

If we observe during a time interval τ and N events are observed, then the average value of N is

$$\overline{N} = \lambda\tau, \quad \text{or} \quad \lambda = \frac{\overline{N}}{\tau} \qquad (3.11)$$

If the observation is repeated many times, the observed value of N will fluctuate around $\overline{N}$. Introducing the fluctuation $\Delta N = (N - \overline{N})$, it will be shown in Chapter 4 that for such events

$$\overline{\Delta N^2} = \operatorname{var} N = \overline{N} = \lambda\tau \qquad (3.12)$$

The inaccuracy of a single measurement is therefore $\Delta\lambda = (\Delta N/\tau)$ and its mean square value is

$$\overline{\Delta\lambda^2} = \overline{\left(\frac{\Delta N}{\tau}\right)^2} = \frac{\overline{\Delta N^2}}{\tau^2} = \frac{\lambda}{\tau} \qquad (3.12a)$$

The r.m.s. inaccuracy of a single measurement thus varies as $1/\tau^{1/2}$. The larger one makes τ, the more accurate the measurement of λ becomes.

We now introduce the limiting accuracy of the measurement by asking what is the smallest value of λ that can be measured. We define it as the value of λ such that

$$\left[\overline{(\Delta\lambda)^2}\right]^{1/2} = \lambda, \quad \text{or} \quad \left(\frac{\lambda}{\tau}\right)^{1/2} = \lambda, \quad \text{or} \quad \lambda\tau = \overline{N} = 1 \qquad (3.12b)$$

The limiting accuracy is thus obtained if on the average one count is made in τ seconds.

As an example consider the counting of electrons. Let the electrons be counted during 5 s. If, on the average, one electron is counted per interval, then the average current is

$$\overline{I} = \frac{e}{\tau} = \frac{1.6 \times 10^{-19}}{5} = 3.2 \times 10^{-20} \text{ A}$$

This is many orders of magnitude better than the measurement of current with the help of galvanometers, indicating the power of the counting technique.

3.2a Current Measurement by Counting Techniques

Current measurements by counting techniques can be achieved if we have free electrons moving in a vacuum. We can then accelerate them and let them impinge upon the first dynode of a secondary emission multiplier with m dynodes (Fig. 3.2). Let each dynode give δ secondary electrons per incident primary; then the output per incident primary is δ^m electrons. This can be made so large that individual pulses can be counted.

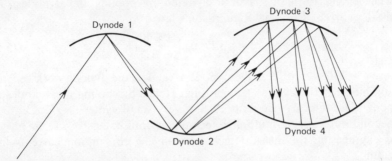

Figure 3.2. Electron multiplication in a secondary emission multiplier. An electron beam is incident upon a first dynode, multiplied, accelerated toward a second dynode, multiplied, acceletated toward a third dynode, multiplied, and so on.

There are two difficulties with this method, as the first dynode: (a) misses an occasional count and (b) emits electrons by thermionic emission. The first difficulty means that the first dynode has a probability p somewhat smaller than unity of producing a pulse when a primary electron strikes it. If the incoming electrons arrive independently and at random, then the pulses are also produced independently and at random. Hence if N electrons come in and n pulses are produced, then

$$\overline{n} = \overline{N}\,p; \qquad \mathrm{var}\,n = \overline{\Delta n^2} = \overline{n} = \overline{N}\,p \qquad (3.13)$$

If the observation time is τ, then the average rate of arrival is $\lambda = (\overline{N}/\tau)$ and the average rate of produced pulses is $(\overline{n}/\tau) = p\lambda$. Therefore,

$$\overline{\left(\frac{\Delta n}{\tau}\right)^2} = \frac{\overline{n}}{\tau^2} = \frac{\overline{N}p}{\tau^2} = \frac{p\lambda}{\tau} \qquad (3.13a)$$

The limiting accuracy is now obtained if

$$\left(\frac{p\lambda}{\tau}\right)^{1/2} = p\lambda \quad \text{or} \quad p\lambda\tau = 1 \quad \text{or} \quad \overline{N} = \frac{1}{p} \qquad (3.13b)$$

In secondary emission multipliers giving a multiplication δ per stage, p is approximately given by (see Chapter 4)

$$p = 1 - \exp(-\delta) \qquad (3.14)$$

so that the effect is insignificant for $\delta > 4$, whereas it is quite important for smaller δ. In addition, some of the primary electrons are reflected by the dynode and do not produce secondary electrons; this is not taken into account by (3.14).

The second difficulty is best obviated by cooling the dynodes. The emission current I follows Richardson's law

$$I = CST^2 \exp\left(-\frac{e\Phi}{kT}\right) \qquad (3.15)$$

where $C = 120$ A/cm^2, S is the dynode area, Φ is the dynode work function in eV, and T is the dynode temperature. Hence by cooling the dynodes the exponential factor can be reduced to a very small value.

Secondary emission-multiplier techniques find extensive use in physics and engineering for measuring very small electron or ion currents in a vacuum.

3.2b Radiation Measurements by Counting Techniques

The same considerations apply to counting photons. If the photons come from a laser, the emission of photons occurs independently and at random, namely $\text{var}\, N = \overline{N}$, where N is the number of emitted photons during a given time interval τ.

In carrying out the counting of the photons one must now use a photomultiplier. It consists of a photocathode emitting photoelectrons that are focused onto the first dynode of a secondary emission multiplier that multiplies the incident electrons and produces sufficiently large electrical pulses that can be counted.

The difficulty with this technique is that the photocathode has generally a low quantum yield η, defined as the average number of photoelectrons emitted per incident quantum. That is the price one has to pay for using photon-counting techniques. The theory of the previous section applies of course, provided that p is replaced by η. This means that the minimum number of photons that can be counted during τ seconds is

$$\overline{N} = \frac{1}{\eta} \qquad (3.16)$$

Example: $\eta = 0.05$, photon energy $V_{ph} = 2.0$ eV; $\tau = 5$ s.

In this case $\overline{N} = (1/\eta) = 20$. The minimum detectable energy is $E_{\min} = e\overline{N}V_{ph} = (1.6 \times 10^{-19}) \times 20 \times 2.0 = 6.4 \times 10^{-18}$ joule and the minimum detectable power is

$$P_{\min} = \frac{E_{\min}}{\tau} = \frac{6.4 \times 10^{-18}}{5} = 1.28 \times 10^{-18} \text{ W}$$

4

TYPICAL DISTRIBUTION FUNCTIONS: VARIANCE THEOREM

In this chapter we discuss three typical distribution functions occurring in physics and engineering: (a) the binomial law, (b) Poisson's law, and (c) the normal law. In addition, we discuss Burgess's variance theorem for calculating averages in complicated statistical problems.

4.1 TYPICAL DISTRIBUTION FUNCTIONS

4.1a Binomial Process

A binomial process is defined as follows. Let an experiment be tried m times. Let p be the probability that the experiment is sucessful and $1-p$ the probability that it fails. Let there be n successes for the m trials, then

$$\overline{n} = mp; \qquad \text{var } n = mp(1-p) \tag{4.1}$$

and the probability distribution is

$$P_m(n) = \frac{m!}{n!(m-n)!} p^n (1-p)^{m-n} \tag{4.2}$$

This is called the binomial law, since $P_m(n)$ is the $(n+1)$th term in the binomial expansion of $[p+(1-p)]^m$.

Before proving these theorems, we first discuss a few cases to which it applies.

1. Vacuum Pentode. "Trial" is the emission of an electron by the cathode, "success" means that it arrives at the anode, and "failure" means that it is intercepted by the screen grid.

2. *Transistor.* "Trial" is the injection of a minority carrier from the emitter into the base, "success" means that it is collected by the collector, and "failure" means that it recombines in the base.

3. *Photodiode.* "Trial" is the absorption of a photon by the $p-n$ junction, "success" is when the generated hole–electron pair is collected, and "failure" means that the pair is not collected.

4. *Photoemissive Cathode.* "Trial" is the absorption of a photon by the cathode, "success" means that a photoelectron is emitted, and "failure" means that no electron is emitted.

5. *Partly Silvered Mirror.* "Trial" is the arrival of a photon at the mirror, "success" means that it is transmitted, and "failure" means that it is reflected.

Next a few words about (4.1). The first part is obvious. Since p is the probability of success and the trials are independent, $\bar{n}$ is equal to the number of trials times the probability of success per trial, as stated. The second part is understandable by bearing in mind that success and failure are interchangeable concepts, and hence $\text{var}\, n$ must be symmetrical in p and $(1-p)$; the second half of (4.1) is the simplest expression that does so. Because individual trials are independent, it is also obvious that $\text{var}\, n$ must be proportional to m; hence it is sufficient to prove the second half of (4.1) for $m=1$. However, that is simple, for if $m=1$, then n is either 0 or 1. Now if $n=0$, then $n^2=0$ and if $n=1$, then $n^2=1$, so that $n^2=n$ in either case. Hence

$$\bar{n}=p; \qquad \overline{n^2}=\bar{n}=p$$

$$\text{var}\, n = \overline{n^2} - (\bar{n})^2 = p - p^2 = p(1-p)$$

as had to be proved.

It should be noted that $\text{var}\, n=0$ for $p=0$ and $p=1$ and that it has its maximum value for $p=\tfrac{1}{2}$, in which case $\text{var}\, n = \tfrac{1}{4}m$.

Next we prove (4.2). The probability that m trials will just give n successes and $m-n$ failures in a prescribed order is $p^n(1-p)^{m-n}$. Yet there are

$$\frac{m!}{n!(m-n)!}$$

ways of assigning the n successes to possible positions in the series of m trials and all are equally probable. Hence (4.2) is correct.

Another important example of binomial statistics is Fermi statistics. In Fermi statistics an "energy state" is described by quantum numbers

Typical Distribution Functions

n_1, n_2, n_3 and m_s, where n_1, n_2, n_3 describe the wave pattern and m_s describes the spin orientation. According to Pauli's exclusion principle an energy state is either empty or occupied by one electron. Hence binomial statistics applies. If f is the probability that the energy state is occupied and n is the number of electrons in that state, then

$$\bar{n} = f; \quad \text{var}\, n = f(1-f) = \bar{n} - (\bar{n})^2 \qquad (4.3)$$

There is nothing mysterious about the latter formula; it is a simple consequence of Pauli's exclusion principle.

4.1b Poisson Process

The Poisson process can be stated as follows. Let a number of events occur independently and at random instants. If n events occur in a particular time interval τ, then

$$\text{var}\, n = \bar{n} \qquad (4.4)$$

and the distribution function $P(n)$ for n is the *Poisson distribution*

$$P(n) = \frac{(\bar{n})^n}{n!} \exp(-\bar{n}) \qquad (4.5)$$

Before proving these theorems, a few examples of such a process will be given.

1. *Saturation Current of a Thermionic Cathode.* The electrons are emitted independently and at random instants. Hence the electron emission is a Poisson process.

2. *Injection of Electrons into the* **P region of an** n^+-p **diode.** The electrons are injected independently and at random instants. Hence electron injection is a Poisson process.

3. *Emission of Photons by a Laser.* The photons are emitted independently and at random instants; hence this emission of photons is a Poisson process. Since the arrival of photons coming from a laser is also a series of independent events occurring at random instants, the arrival of photons coming from a laser is also a Poisson process.

However, the emission of black-body radiation is not a Poisson process, since the photons are emitted in bunches. Consequently, $\text{var}\, n > \bar{n}$; the emission process is then said to be *superpoissonian*.

To prove (4.4) and (4.5) we divide the interval τ into m time intervals of equal length τ/m, where $m \gg \bar{n}$. Then the probability $p = (\bar{n}/m)$ that a single event occurs during a time interval τ/m is very small and the probability that more than one event occurs is negligible. Therefore, the binomial law holds, or

$$\bar{n} = mp; \quad \operatorname{var} n = mp(1-p)$$

and the distribution function is $P_m(n)$. We now let m go to infinity such that $mp = \bar{n}$ remains constant. Then $p \to 0$ and hence $\operatorname{var} n = mp = \bar{n}$, as had to be proved.

To prove (4.5) we must prove that

$$\lim_{m \to \infty, mp = \bar{n}} P_m(n) = \frac{(\bar{n})^n}{n!} \exp(-\bar{n})$$

The proof is simple. We write

$$P_m(n) = \frac{(1-p)^m}{n!} \frac{m!}{(m-n)!} (1-p)^{-n} p^n$$

Now

$$(1-p)^m = \left[(1-p)^{1/p}\right]^{mp} = \left[(1-p)^{1/p}\right]^{\bar{n}} = \left(\frac{1}{e}\right)^{\bar{n}} = \exp(-\bar{n})$$

since $\lim_{p \to 0}(1-p)^{1/p} = 1/e$. Furthermore $\lim_{p \to 0}(1-p)^{-n} = 1$ for any finite n. Therefore we must prove

$$\lim_{m \to \infty, mp = \bar{n}} \frac{m! p^n}{(m-n)!} = (\bar{n})^n$$

Now since $p = (\bar{n}/m)$,

$$\lim_{m \to \infty, mp = \bar{n}} \frac{m!}{(m-n)!} p^n = \lim_{m \to \infty} \frac{m!}{(m-n)!} \frac{(\bar{n})^n}{m^n}$$

$$\lim_{m \to \infty} \frac{m}{m} \cdot \frac{m-1}{m} \cdot \frac{m-2}{m} \cdots \frac{m-n+1}{m} \cdot (\bar{n})^n = (\bar{n})^n$$

as had to be proved.

We now apply Poisson's law to the secondary emission dynode to evaluate the probability that no secondary electron is emitted. We assume that the secondary multiplication process may approximately be represented by a Poisson process as long as the primary energy is not too large (see Section 4.2). Hence the probability that no electron is emitted is

$$P(0) = 1 - p = \frac{(\bar{n})^0}{0!} \exp(-\bar{n}) = \exp(-\bar{n}) = \exp(-\delta)$$

or $p = 1 - P(0) = 1 - \exp(-\delta)$. This proves (3.14).

4.1c The Normal Law

For large values of n the binomial distribution $P_m(n)$ and the Poisson distribution $P(n)$ can be approximated by the so-called *normal distribution*

$$P(n) = \frac{1}{(2\pi \operatorname{var} n)^{1/2}} \exp\left[-\frac{(n-\bar{n})^2}{2 \operatorname{var} n}\right] \qquad (4.6)$$

The case $\operatorname{var} n = \bar{n}$ is sometimes called the *Gaussian* distribution. For the Poisson (or shot noise) process, $\operatorname{var} n = \bar{n}$.

The proof of (4.6) proceeds as follows. Let $P_m(n)$ have a maximum value at $n = n_0 \cong \bar{n}$. We now make a Taylor expansion of $\ln P_m(n)$ with respect to $\Delta n = (n - n_0)$ and terminate with the second-order term; this is sufficiently accurate if n_0 is large. Since n is a large number, we may define

$$\frac{d}{dn}\left[\ln P_m(n)\right] \cong \frac{\ln P_m(n+1) - \ln P_m(n)}{n+1-n} \qquad (4.6a)$$

Setting this equal to zero, we find n_0. Evaluating the second derivative at $n = n_0$ by differentiation, we have

$$\ln P_m(n) = \ln P_m(n_0) + \frac{1}{2} \frac{d^2}{dn^2} \ln P_m(n)\bigg|_{n=n_0} \Delta n^2 \qquad (4.6b)$$

Comparing this to (4.6) shows complete identity and yields $\operatorname{var} n$.

We can have a similar distribution for a continuous random variable x with average value $\bar{x}$ and variance $\sigma_x^2 = \overline{x^2} - (\bar{x})^2$. In analogy with (4.6)

$$P(x) = \frac{1}{(2\pi \sigma_x^2)^{1/2}} \exp\left[-\frac{(x-\bar{x})^2}{2\sigma_x^2}\right] \qquad (4.7)$$

This is easily extended to the case of more variables. In the case of two variables x and y one must consider the averages $\overline{x^2}$, $\overline{xy}$, and $\overline{y^2}$. We first discuss the case $\overline{xy} = 0$. The two variables are then independent and one has the joint-probability density function

$$P(x,y) = \frac{1}{2\pi(\sigma_x^2\sigma_y^2)^{1/2}} \exp\left[-\frac{(x-\overline{x})^2}{2\sigma_x^2} - \frac{(y-\overline{y})^2}{2\sigma_y^2}\right] \quad (4.8)$$

One can now make an orthogonal transformation to new random variables x_1 and x_2 that may be partially correlated. One then obtains*

$$P(x_1, x_2) = \frac{1}{2\pi M^{1/2}} \exp\left[-\tfrac{1}{2}(\mathbf{x}'\mathbf{M}^{-1}\mathbf{x})\right] \quad (4.9)$$

where x, x', and $\mathbf{M}$ are matrices

$$\mathbf{x} = \begin{Bmatrix} x_1 \\ x_2 \end{Bmatrix}; \quad \mathbf{x}' = \{x_1, x_2\}; \quad \mathbf{M} = \begin{Bmatrix} \overline{x_1^2} & \overline{x_1 x_2} \\ \overline{x_1 x_2} & \overline{x_2^2} \end{Bmatrix} \quad (4.9a)$$

$\mathbf{M}^{-1}$ is the reciprocal of the matrix $\mathbf{M}$ and M is the determinant of $\mathbf{M}$.

This is easily extended to more variables.*

It turns out that practically all noise phenomena encountered in physics and engineering can be described by the normal distribution. This is a consequence of the *central limit theorem*, which holds for practically all noise phenomena. It can be formulated as follows:†

If $X_1, X_2 \ldots X_n$ are independent random variables, all having the same probability density function, and hence equal averages $\overline{X_i} = \overline{X_1}$ and equal variances $\operatorname{var} X_i = \operatorname{var} X_1 = \sigma_1^2$, then the sum $Y = \sum_{i=1}^{n} X_i$ is asymptotically normal for large n, with an average $\overline{Y} = n\overline{X_1}$ and variance $\sigma_1^2 n$, if σ_1 exists.

As a consequence very little information can be gained by measuring the distribution function for practically all noise phenomena, since it will most likely be the normal distribution anyway.

4.2 THE VARIANCE THEOREM

There is a large class of noise problems that can be described by the following process, identified by Burgess.*

Let a sequence of N events occur during a time interval τ. Let a quantity a_i be assigned to each event. Let a second quantity n be defined by

$$n = \sum_{i=1}^{N} a_i \quad (4.10)$$

---

*A. van der Ziel, *Noise*, Prentice Hall, Englewood Cliffs, N. J., 1954, Appendix 1.
†H. Cramér, *Mathematical Methods of Statistics*, Princeton U. P., 1946.
*R. E. Burgess, *Faraday Soc. Disc.* **28**, 151 (1959).

The Variance Theorem

Let now N and a_i both fluctuate and let $\overline{a_i} = \overline{a}$ and $\overline{a_i^2} = \overline{a^2}$ be independent of i. Then

$$\overline{n} = \overline{N}\,\overline{a} \ ; \qquad \mathrm{var}\, n = (\overline{a})^2 \mathrm{var}\, N + \overline{N}\, \mathrm{var}\, a \tag{4.11}$$

(Burgess's variance theorem).

To prove the variance theorem we start with an ensemble and average over it in three steps:

1. We divide the ensemble into subensembles with the same value of N.
2. We first average over each subensemble separately.
3. We finally average over all subensembles.

We shall denote a subensemble average by $\overline{}^s$ and the ensemble average by $\overline{}$. We then have $\overline{n} = \overline{\overline{n}^s} = \overline{N\overline{a}} = \overline{N}\,\overline{a}$, which proves the first half of (4.11). To prove the second half we observe that

$$n^2 = \sum_{i=1}^{N} \sum_{j=1}^{N} a_i a_j$$

Hence

$$\overline{n^2}^s = \sum_{i=1}^{N} \sum_{j=1}^{N} \overline{a_i a_j}^s = N(N-1)(\overline{a})^2 + N\overline{a^2}$$

since there are N terms $a_i a_j$ with $i = j$ for which $\overline{a_i a_j}^s = \overline{a^2}$ and $N(N-1)$ terms $a_i a_j$ with $i \neq j$ for which $\overline{a_i a_j}^s = (\overline{a})^2$. Hence

$$\overline{n^2}^s = N^2 (\overline{a})^2 + N\, \mathrm{var}\, a \quad \text{or}$$

$$\overline{n^2} = \overline{\overline{n^2}^s} = \overline{N^2}(\overline{a})^2 + \overline{N}\, \mathrm{var}\, a$$

so that

$$\mathrm{var}\, n = \overline{n^2} - (\overline{n})^2 = (\overline{a})^2 \left[\overline{N^2} - (\overline{N})^2 \right] + \overline{N}\, \mathrm{var}\, a$$

which proves the second half of (4.11).

4.2a Applications to Pentodes and Transistors: Partition Noise

As a first example we consider the *vacuum pentode*. Let n_c electrons be emitted by the cathode; let n_a arrive at the anode and n_2 at the screen grid. Under space-charge limited conditions in the cathode-grid region, the cathode-current fluctuations are smoothed by the space charge in front of the cathode, resulting in $\text{var}\, n_c < n_c$.

For this case n_c corresponds to N. Let $a_i = 1$ if the electron arrives at the anode and $a_i = 0$ if the electron arrives at the screen grid. If λ is the probability that the electron arrives at the anode, then $\overline{a} = \lambda$ and $\text{var}\, a = \lambda(1-\lambda)$. Hence the variance theorem applies and

$$\overline{n_a} = \overline{n_c}\lambda; \qquad \overline{n_2} = \overline{n_c}(1-\lambda) \qquad (4.12)$$

$$\text{var}\, n_a = \lambda^2 \text{var}\, n_c + \overline{n_c}\lambda(1-\lambda); \qquad \text{var}\, n_2 = (1-\lambda)^2 \text{var}\, n_c + \overline{n_c}\lambda(1-\lambda) \qquad (4.13)$$

The term $\lambda^2 \text{var}\, n_c$ is called *attenuated shot noise* and the term $\overline{n_c}\lambda(1-\lambda)$ is called *partition noise*.

We can rewrite the first half of (4.13) as

$$\text{var}\, n_a = \overline{n_c}\lambda + \lambda^2(\text{var}\, n_c - \overline{n_c}) = \overline{n_a} + \lambda^2(\text{var}\, n_c - \overline{n_c}) \qquad (4.13\text{a})$$

Hence if the cathode noise is not attenuated by space charge, or $\text{var}\, n_c = \overline{n_c}$, then $\text{var}\, n_a = \overline{n_a}$. In other words the anode current has full-shot noise if the cathode current has full-shot noise.

We can also write

$$n_c = n_a + n_2 \qquad \Delta n_c = \Delta n_a + \Delta n_2$$

$$\overline{\Delta n_c^2} = \text{var}\, n_c = \overline{\Delta n_a^2} + 2\overline{\Delta n_a \Delta n_2} + \overline{\Delta n_2^2}$$

$$= \text{var}\, n_a + \text{var}\, n_2 + 2\overline{\Delta n_a \Delta n_2}$$

$$= [\lambda^2 + (1-\lambda)^2]\text{var}\, n_c + 2\overline{n_c}\lambda(1-\lambda) + 2\overline{\Delta n_a \Delta n_2}$$

Hence since $1 - \lambda^2 - (1-\lambda)^2 = 2\lambda(1-\lambda)$

$$2\overline{\Delta n_a \Delta n_2} = -2\overline{n_c}\lambda(1-\lambda) + 2\lambda(1-\lambda)\text{var}\, n_c$$

or

$$\overline{\Delta n_a \Delta n_2} = (\text{var}\, n_c - \overline{n_c})\lambda(1-\lambda) \qquad (4.14)$$

Therefore, if $\text{var}\, n_c = \overline{n_c}$, Δn_a, and Δn_2 are independent.

We can also write this as follows: Put

$$\Delta n_a = \lambda \Delta n_c + \Delta n_p; \qquad \Delta n_2 = (1-\lambda)\Delta n_c - \Delta n_p$$

where Δn_p is independent of Δn_c. This set of equations makes sense, for an

The Variance Theorem

electron going to the screen grid is missed at the anode and vice versa. Hence according to (4.13)

$$\operatorname{var} n_a = \lambda^2 \overline{\Delta n_c^2} + \overline{\Delta n_p^2}, \quad \text{or} \quad \overline{\Delta n_p^2} = \overline{n_c} \lambda(1-\lambda) \qquad (4.15)$$

and

$$\overline{\Delta n_a \Delta n_2} = \lambda(1-\lambda)\overline{\Delta n_c^2} - \overline{\Delta n_p^2} = \lambda(1-\lambda)(\operatorname{var} n_c - \overline{n_c})$$

in agreement with (4.14). The term $\overline{\Delta n_p^2}$ is called *partition noise*; it is the only noise observed when the cathode current fluctuations are fully smoothed out.

As a second example we consider the *transistor*. Let n_E be the rate of electron injection into the base by the emitter; n_B the rate of recombination in the base; and n_C the rate of electron collection by the collector. Then the variance theorem holds and

$$n_E = n_B + n_C; \quad \overline{n_B} = (1-\alpha_F)\overline{n_E}; \quad \overline{n_C} = \alpha_F \overline{n_E} \qquad (4.16)$$

where α_F is the d.c. current amplification factor of the transistor. Moreover, since $\operatorname{var} n_E = \overline{n_E}$,

$$\operatorname{var} n_C = \alpha_F^2 \operatorname{var} n_E + \overline{n_E} \alpha_F (1-\alpha_F) = \overline{n_E} \alpha_F = \overline{n_C} \qquad (4.17)$$

$$\operatorname{var} n_B = (1-\alpha_F)^2 \operatorname{var} n_E + \overline{n_E} \alpha_F (1-\alpha_F) = \overline{n_E}(1-\alpha_F) = \overline{n_B} \qquad (4.18)$$

Finally, if $\Delta n_B = n_B - \overline{n_B}$ and $\Delta n_C = n_C - \overline{n_C}$,

$$\overline{\Delta n_B \Delta n_C} = (\operatorname{var} n_E - \overline{n_E})\alpha_F(1-\alpha_F) = 0 \qquad (4.19)$$

in analogy with (4.14). In other words, in a transistor the collector and the base currents fluctuate independently and each current shows full shot noise.

4.2b Electron-Multiplication Processes

We now apply our results to an electron-multiplication process. Let N be the number of electrons arriving at the device and let $\operatorname{var} N = \overline{N}$ (Poisson process). Let a_i be the number of secondary electrons produced by the ith primary; since all primaries have equal probability as far as multiplication is concerned, $\overline{a_i} = \overline{a}$ and $\overline{a_i^2} = \overline{a^2}$, independent of i. Hence the variance theorem is valid and

$$\overline{n} = \overline{N}\,\overline{a}; \quad \operatorname{var} n = (\overline{a})^2 \operatorname{var} N + \overline{N} \operatorname{var} a \qquad (4.20)$$

The first term in var n is called *amplified primary noise* and the term $\overline{N}$ var a is called *multiplication noise*. If we now make use of the fact that var $N = \overline{N}$ we obtain

$$\text{var}\, n = \overline{N}\, \overline{a^2} = \overline{N}\, \overline{a}\, \frac{\overline{a^2}}{\overline{a}} = \overline{n}\, \frac{\overline{a^2}}{\overline{a}} \tag{4.20a}$$

In the particular case that var $a = \overline{a}$, we have $\overline{a^2} = \overline{a} + (\overline{a})^2$. Then

$$\text{var}\, n = \overline{n}\,(\overline{a} + 1) \tag{4.20b}$$

In the case that by some method var N has been completely smoothed out, so that var $N = 0$, only the multiplication noise is left, or

$$\text{var}\, n = \overline{N}\, \text{var}\, a \tag{4.20c}$$

We now apply this to hole–electron pair production in a $p-n$ junction, or to hole–electron pair production in a photoconductive cell, by high-energy quanta, electrons, or heavier particles. In that case the preceding theory is valid, and if var $N = \overline{N}$ the noise is completely described by (4.20a).

We now consider the case of hole–electron pair production in $p-n$ junctions, in which the pairs have a probability λ of being collected. If m pairs are collected for n pairs produced, then, according to the variance theorem

$$\overline{m} = \overline{n}\lambda = \overline{N}\,\overline{a}\lambda \tag{4.21}$$

$$\text{var}\, m = \lambda^2 \text{var}\, n + \overline{n}\lambda(1-\lambda) = \overline{n}\lambda^2 \frac{\overline{a^2}}{\overline{a}} + \overline{n}\lambda(1-\lambda)$$

$$= \overline{m}\left(\lambda \frac{\overline{a^2}}{\overline{a}} + 1 - \lambda\right) \tag{4.22}$$

In the particular case that the multiplication process is also a Poisson process, we have var $a = \overline{a} = \overline{a^2} - (\overline{a})^2$, or $\overline{a^2} = (\overline{a})^2 + \overline{a}$. Consequently, from (4.22)

$$\text{var}\, m = \overline{m}\,(\lambda \overline{a} + 1) \tag{4.23}$$

Something should be said about the escape probability of produced photoelectrons in photocathodes.* Formerly this escape probability was

---

*A. van der Ziel, *Solid State Physical Electronics*, 3rd ed., Prentice Hall, Englewood Cliffs, N. J., 1976.

not very high because the electron affinity χ of the semiconductor surface was positive (Fig. 4.1a). Since only those photoelectrons can escape for which $\frac{1}{2}mv_x^2 > \chi$, where v_x is the velocity component perpendicular to the surface, the escape probability of the photoelectrons was quite small. Photocathodes with a negative electron affinity χ exist; in that case potentially all electrons that are brought into the conduction band can escape (Fig. 4.1b). The electrons can now lose their energy by pair production if $\frac{1}{2}mv^2 > eE_g$ and to lattice vibrations if $\frac{1}{2}mv^2 < eE_g$; here v is the velocity of the photoelectrons.

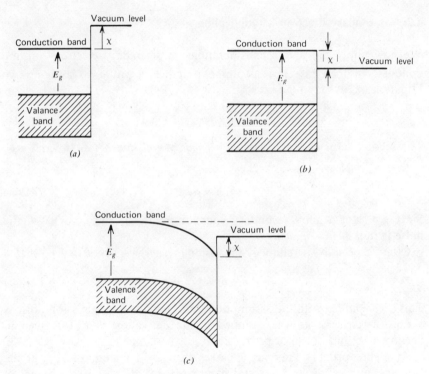

Figure 4.1. (a) Semiconductor band diagram with valence band, conduction band, gap width E_g, and positive electron affinity χ;(b) same but with negative electrons affinity χ;(c) same but with positive electron affinity χ and band bending near the surface, so that the bottom of the conduction band in the bulk is above the vacuum level.

To make the thermionic emission of the cathode as small as possible, one should make the work function of the semiconductor surface as large as possible. This is achieved by making the photocathode p^+ type. The photoelectrons along their way to the surface will now recombine with holes. If the electrons have a lifetime τ_n and a diffusion constant D_n, then their diffusion length $L_n = (D_n \tau_n)^{1/2}$. The travel to the surface is a diffusion process, and the escape depth of the photoelectrons equals the diffusion length L_n. This can be as large as several μm.

The case pictured in Fig. 4.1b is the so-called *flatband case*, where there is no band bending at the surface. This is the case for GaAs and GaP surfaces covered by an —O—Cs layer. In other cases, as in $In_xGa_{1-x}As$ or Si surfaces covered with an —O—Cs layer, there is band bending at the surface (Fig. 4.1c); the escape probability is now smaller than in the flatband case.

4.2c Secondary Electron Multiplication

We now apply the above considerations to secondary electron-emitting dynodes. In that case it is common practice to put $\bar{a} = \delta$ and $\overline{a^2} = \kappa \delta$. Equation (4.20) may now be written

$$\bar{n} = \bar{N}\delta; \qquad \text{var}\, n = \delta^2 \text{var}\, N + \bar{N}\delta(\kappa - \delta) \qquad (4.24)$$

The second term is usually called *secondary emission noise*. If $\text{var}\, N = \bar{N}$, then

$$\text{var}\, n = \bar{n}\kappa \qquad (4.24a)$$

Since $\kappa \gg 1$ for a good secondary emission-multiplier stage, $\text{var}\, n \gg \bar{n}$; the noise is thus *super-Poissonian*.

Often, especially at relatively low primary energies, $(\kappa - \delta) \simeq 1$ so that

$$\text{var}\, n = \delta^2 \text{var}\, N + \bar{n} \qquad (4.24b)$$

This is *compatible* with the assumption that the secondary emission noise is a Poisson process, as was assumed in the derivation of (3.14) given in Section 4.1b, but does not prove it.

At higher primary energies δ passes through a maximum δ_{max} at the primary energy $E_{p0} = E_{max}$ and decreases for $E_{p0} > E_{max}$, whereas κ saturates beyond E_{max}.* If E_{p0} is so chosen that κ/δ is a minimum, the relative importance of the secondary emission noise is minimized.

The best secondary emitters are GaAs and GaP, covered with an O—Cs monolayer; they have rather high δ values at relatively low primary energies and a very high δ_{max} at a relatively high primary energy E_{max}. As mentioned before, these emitters should be of the heavily doped p type to increase the work function and hence lower the thermionic emission.

*C. B. Murray, *Physica*, **38**, 549 (1968).

4.2d The Fano Factor

In many high-energy particle or quantum detection systems one uses multiplication processes that produce a large number of electrons or hole–electron pairs for each individual incoming particle or quantum. One can then characterize the multiplication process by the average number $\bar{a}$ of particles produced by each elementary event and by the variance in that number, $\mathrm{var}\,a$.

If the multiplication process were a Poisson process, we would have $\mathrm{var}\,a = \bar{a}$; since the multiplication process must satisfy the constraint that the *fixed* energy E of the incoming particle or quantum is used to produce electrons or hole–electron pairs, $\mathrm{var}\,a$ may differ from $\bar{a}$. It is then common practice to introduce the Fano factor F of the multiplication process, defined by the relation

$$\mathrm{var}\,a = F\bar{a} \tag{4.25}$$

Usually $\bar{a}$ is proportional to the energy E of the incoming particles or quanta, and hence the height of the pulses due to individual events is a measure for the energy E. Since $\mathrm{var}\,a$ corresponds to a spread in the pulse height, it characterizes an uncertainty in the measured energy E. If one wants to resolve current pulses due to particles or quanta of slightly different energies, then $\mathrm{var}\,a$ and hence the Fano factor F should be as small as possible.

This book discusses a variety of multiplication processes in subsequent chapters. The Fano factors of these processes are dealt with in Chapter 17.

5

FOURIER ANALYSIS OF FLUCTUATING QUANTITIES: SPECTRAL INTENSITIES

In this chapter we make a Fourier analysis of a random noise signal $X(t)$ and define the spectral intensity $S_x(f)$ of $X(t)$. We then extend it to partly correlated random variables $X(t)$ and $Y(t)$ and introduce self-spectral intensities $S_{xx}(f)$ and $S_{yy}(f)$ and cross-spectral intensities $S_{xy}(f)$ and $S_{yx}(f)$ (Section 5.1).

In Section 5.2 we discuss a wide variety of methods for calculating spectral intensities.

5.1 FOURIER ANALYSIS; WIENER–KHINTCHINE THEOREM

To find the response of a measuring system with a frequency-response function $g(f)$ to a noise signal $X(t)$, it is necessary to make a Fourier analysis of $X(t)$. To that end we must apply the *Fourier theorem*, which can be formulated as follows:

Let a piecewise continuous function $X(t)$ be defined for the time interval $0 < t < T$, and let at a discontinuity at $t = t_0$ the function $X(t)$ have the value

$$X(t_0) = \lim_{h>0, h\to 0} \tfrac{1}{2}\left[X(t_0+h) + X(t_0-h)\right]$$

Let at $t = 0$ and $t = T$ the function $X(t)$ have the value

$$X(0) = X(T) = \lim_{h>0, h\to 0} \tfrac{1}{2}\left[X(h) + X(T-h)\right]$$

Fourier Analysis; Wiener–Khintchine Theorem

Then $X(t)$ is equal to its Fourier series

$$X(t) = \sum_{n=-\infty}^{\infty} a_n \exp(j\omega_n t) \tag{5.1}$$

for $0 \leq t \leq T$, endpoints and discontinuities included, where $\omega_n = (2\pi n/T)$ $(n = 0, \pm 1, \pm 2, \ldots)$ and

$$a_n = \frac{1}{T} \int_0^T X(t) \exp(-j\omega_n t) \, dt \tag{5.1a}$$

It is obvious that, if a_n^* is the complex conjugate of a_n, then

$$a_{-n} = a_n^* \tag{5.1b}$$

We further observe that for a given T the Fourier component of frequency ω_n is

$$x_n = a_n \exp(j\omega_n t) + a_n \exp(-j\omega_n t) \tag{5.2}$$

If we apply this to the elements of an ensemble of identical systems subjected to independent fluctuations, then the a_n values have arbitrary phase. That is, if we take ensemble averages

$$\overline{x_n^2} = \overline{a_n^2} \exp(j\omega_n t) + \overline{a_{-n}^2} \exp(-j\omega_n t) + 2\overline{a_n a_n^*} = 2\overline{a_n a_n^*} \tag{5.3}$$

since $\overline{a_n^2} = \overline{a_{-n}^2} = 0$ because of the arbitrary phase. This is the starting point of defining spectral intensities.

The above considerations are valid both for $\overline{X} = 0$ and $\overline{X} \neq 0$. Since it is always possible to introduce $X(t) - \overline{X}$ as the fluctuating quantity, we may assume without loss of generality that $\overline{X} = 0$.

5.1a Wiener–Khintchine Theorem; Spectral Intensity

We now evaluate $2\overline{a_n a_n^*}$. To that end we observe that

$$a_n = \frac{1}{T} \int_0^T X(u) \exp(-j\omega_n u) \, du; \quad a_n^* = \frac{1}{T} \int_0^T X(v) \exp(j\omega_n v) \, dv$$

so that

$$2\overline{a_n a_n^*} = \frac{2}{T^2} \int_0^T \int_0^T \overline{X(u)X(v)} \exp[j\omega_n(v-u)] \, du \, dv \tag{5.4}$$

The integrand has the largest value $\overline{X^2(v)}$ along the line $u=v$ and decreases more or less rapidly away from this diagonal (Fig. 5.1). We now introduce another domain of integration by drawing two lines parallel to the diagonal at a vertical distance $\pm M$ from the diagonal. We choose M so large that if $s=(v-u)$

$$\int_{-M}^{M} \overline{X(u)X(u+s)} \exp(j\omega_n s)\,ds \simeq \int_{-\infty}^{\infty} \overline{X(u)X(u+s)} \exp(j\omega_n s)\,ds \quad (5.5)$$

This can be done if the integral converges absolutely. We now choose $T \gg M$. Then

$$\frac{2}{T^2}\int_0^T du \int_{-u}^{T-u} \overline{X(u)X(u+s)} \exp(j\omega_n s)\,ds$$

$$\simeq \frac{2}{T^2}\int_0^T du \int_{-M}^{M} \overline{X(u)X(u+s)} \exp(j\omega_n s)\,ds \quad (5.6)$$

since the two areas of integration differ only over the two large triangles for which $\overline{X(u)X(u+s)}=0$ and over the two small triangles of side length M; their contribution is negligible if $T \gg M$. Now

$$\int_{-M}^{M} \overline{X(u)X(u+s)} \exp(j\omega_n s)\,ds$$

is independent of u (stationarity) and $\int_0^T du = T$. Hence

$$2\overline{a_n a_n^*} \simeq \frac{2}{T}\int_{-M}^{M} \overline{X(u)X(u+s)} \exp(j\omega_n s)\,ds$$

$$\simeq \frac{2}{T}\int_{-\infty}^{\infty} \overline{X(u)X(u+s)} \exp(j\omega_n s)\,ds \quad (5.7)$$

since the integral converges absolutely. Therefore if $\Delta f = 1/T$, the frequency interval between frequencies f_n

$$\overline{x_n^2} = S_x(f_n)\Delta f \quad (5.7a)$$

where

$$S_x(f_n) = 2\int_{-\infty}^{\infty} \overline{X(u)X(u+s)} \exp(j\omega_n s)\,ds \quad (5.7b)$$

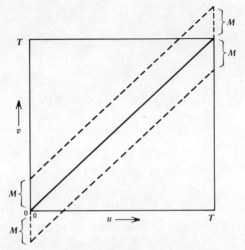

Figure 5.1. Two areas of integration are considered: (*a*) square area of integration of side length T; (*b*) parallelogram of height $2M$. The two areas differ by two large triangles for which the integrand is negligible, and by two small triangles that give a negligible contribution if $T \gg M$.

$S_x(f_n)$ is thus twice the Fourier transform of $\overline{X(u)X(u+s)}$; it is called the *spectral intensity* of $X(t)$. Equation (5.7b) is called the *Wiener–Khintchine theorem*.

When is this derivation valid? A sufficient condition for it is that $\overline{X(u)X(u+s)}$ is absolutely integrable, namely that

$$\int_{-\infty}^{\infty} |\overline{X(u)X(u+s)}|\, ds$$

exists. For if that condition is satisfied the Fourier transform of $\overline{X(u)X(u+s)}$ exists and (5.5) is valid.

Next we prove that

$$\overline{a_n a_m^*} = 0 \quad \text{for} \quad m \neq n \tag{5.8}$$

that is, the Fourier components of different frequencies are independent. The proof is simple. Since

$$a_n = \frac{1}{T}\int_0^T X(u)\exp(-j\omega_n u)\, du; \qquad a_m^* = \frac{1}{T}\int_0^T X(v)\exp(j\omega_m v)\, dv$$

Therefore, if $T \gg M$, using the same procedure as before,

$$\overline{a_n a_m^*} = \frac{1}{T^2} \int_0^T \int_0^T \overline{X(u)X(v)} \exp(-j\omega_n u + j\omega_m v) \, du \, dv$$

$$\cong \frac{1}{T^2} \int_0^T \exp[-j(\omega_n - \omega_m)u] \, du \int_{-M}^{M} \overline{X(u)X(u+s)} \exp(j\omega_m s) \, ds = 0$$

because the first integral is zero for $m \neq n$ and the second integral is finite; here $s = (v - u)$. This proves (5.8).

If we drop the subscript n, we thus have proved the following:

1. It is possible to introduce a spectral intensity $S_x(f)$ of $X(t)$ so that $[S_x(f)\Delta f]^{1/2}$ represents the r.m.s. Fourier component in the frequency interval Δf around f.
2. Fourier components of different frequency intervals are independent.
3. Alternating current circuit analysis can be applied to these r.m.s. components.

Noise calculations thus amount to calculating $S_x(f)$.

5.1b Spectral Intensity Theorems

The spectral intensity $S_x(f)$ obeys the following theorems:

1. If a noise signal $X(t)$ has a spectral intensity $S_x(f)$, then

$$\overline{X^2} = \int_0^\infty S_x(f) \, df \tag{5.9}$$

In other words, if we know $S_x(f)$, we know $\overline{X^2}$.

Proof. Since $X(t) = \sum_{n=-\infty}^{\infty} a_n \exp(j\omega_n t)$

$$\overline{X^2} = \sum_{n=-\infty}^{\infty} \overline{a_n a_{-n}} = \sum_{n=0}^{\infty} 2\overline{a_n a_n^*} = \sum_{n=0}^{\infty} S_x(f_n) \Delta f_n = \int_0^\infty S_x(f) \, df$$

since $\overline{a_n a_m^*} = 0$ for $m \neq n$ and $a_{-n} = a_n^*$.

2.

$$S_x(f) = 2 \int_{-\infty}^{\infty} \overline{X(t)X(t+s)} \exp(j\omega s) \, ds = 4 \int_0^\infty \overline{X(t)X(t+s)} \cos \omega s \, ds$$

$$\tag{5.10}$$

Proof. $\exp(j\omega s) = \cos\omega s + j\sin\omega s$; since $\overline{X(t)X(t+s)}$ and $\cos\omega s$ are symmetric in s whereas $\sin\omega s$ is antisymmetric in s, we have

$$S_x(f) = 2\int_{-\infty}^{\infty} \overline{X(t)X(t+s)} \cos\omega s\, ds = 4\int_{0}^{\infty} \overline{X(t)X(t+s)} \cos\omega s\, ds$$

This relationship should be used with caution if $\overline{X(t)X(t+s)}$ is a delta function in s.

3. If (5.10) is valid, then

$$\overline{X(t)X(t+s)} = \int_0^{\infty} S_x(f)\cos\omega s\, df \tag{5.11}$$

This follows simply from inverting (5.10).

4. For $\overline{X^2}$ to exist, (5.9) must converge. This means that two frequencies f_1 and f_2 can be chosen such that $S_x(f)$ varies at a slower rate than $1/f$ for $f<f_1$ and that $S_x(f)$ varies at a faster rate than $1/f$ for $f>f_2$. This relationship is not valid for flicker noise, which has a $1/f^\alpha$ spectrum with $\alpha \simeq 1$; this is an indication that a $1/f$ spectrum cannot extend to the frequency $f=0$ or the frequency $f=\infty$.

5. If a noise signal $X(t)$ with a spectral intensity $S_x(f)$ is applied to an amplifier of (complex) voltage gain $g(f)$, then the output noise signal $Y(t)$ has a mean square value

$$\overline{Y^2} = \int_0^{\infty} S_x(f)|g(f)|^2 df \tag{5.12}$$

Proof. Put $X(t) = \sum_{n=-\infty}^{\infty} a_n \exp(j\omega_n t)$, $Y(t) = \sum_{n=-\infty}^{\infty} b_n \exp(j\omega_n t)$. According to a.c. analysis $b_n = a_n g(f_n)$ so that

$$S_y(f_n) = |g(f_n)|^2 S_x(f_n)$$

Dropping the subscript n and applying (5.9) yields

$$\overline{Y^2} = \int_0^{\infty} S_y(f)\, df = \int_0^{\infty} S_x(f)|g(f)|^2 df$$

6. Equation (5.12) can be used to measure $S_x(f)$. Let the amplifier be sharply tuned at the frequency f_0, that is, the passband B of the amplifier $\ll f_0$. Then $S_x(f) \simeq S_x(f_0)$ for all frequencies of practical interest. If $g(f_0) = g_0$, we have

$$\overline{Y^2} = S_x(f_0) \int_0^{\infty} |g(f)|^2 df = S_x(f_0) g_0^2 B_{\text{eff}} \tag{5.13}$$

where

$$B_{\text{eff}} = \int_0^\infty \left|\frac{g(f)}{g_0}\right|^2 df \qquad (5.13a)$$

Since $\overline{Y^2}$, g_0^2, and B_{eff} can be measured, $S_x(f_0)$ can be determined; B_{eff} is called the *effective bandwidth* of the amplifier.

7. The low-frequency value $S_x(0)$ of $S_x(f)$ may usually be written as

$$S_x(0) = 2\int_{-\infty}^\infty \overline{X(t)X(t+s)}\, ds \qquad (5.14)$$

For if ω is sufficiently small, $\cos\omega s \simeq 1$ for all values of s for which $\overline{X(t)X(t+s)}$ differs from zero. The only exception is again flicker noise, which has a $1/f$ spectrum. For practically all other spectra $S_x(f)$ is independent of frequency at low frequencies. The spectrum is then said to be *white* at those frequencies.

8. If $S_x(f)$ is white at low frequencies, and

$$X_\tau = \frac{1}{\tau}\int_0^\tau X(u)\, du, \quad \text{then} \quad \lim_{\tau\to\infty} 2\tau\, \overline{X_\tau^2} = S_x(0) \qquad (5.15)$$

Proof. If τ is sufficiently large and $S_x(f)$ is white at low frequencies

$$\overline{X_\tau^2} = \frac{1}{\tau^2}\int_0^\tau\int_0^\tau \overline{X(u)X(v)}\, du\, dv \simeq \frac{1}{\tau^2}\int_0^\tau du \int_{-M}^M \overline{X(u)X(u+s)}\, ds$$

$$\simeq \frac{1}{\tau}\int_{-\infty}^\infty \overline{X(u)X(u+s)}\, ds = \frac{S_x(0)}{2\tau}$$

This proves (5.15). Often one can evaluate $\overline{X_\tau^2}$ and hence determine $S_x(0)$ with the help of (5.15).

5.1c Extension to More Random Variables

We now repeat the process for two random variables $X(t)$ and $Y(t)$ that are partially correlated. We have then two autocorrelation functions $\overline{X(t)X(t+s)}$ and $\overline{Y(t)Y(t+s)}$ and two cross-correlation functions $\overline{X(t)Y(t+s)}$ and $\overline{Y(t)X(t+s)}$. Consequently there are two self-spectral intensities $S_{x,x}(f)$ and $S_{y,y}(f)$ and two cross-spectral intensities $S_{x,y}(f)$ and

$S_{y,x}(f)$. Going through the same reasoning as for one variable,

$$S_{x,x}(f) = 2\int_{-\infty}^{\infty} \overline{X(t)X(t+s)} \exp(j\omega s)\, ds;$$

$$S_{y,y}(f) = 2\int_{-\infty}^{\infty} \overline{Y(t)Y(t+s)} \exp(j\omega s)\, ds \qquad (5.16)$$

and, by analogy

$$S_{x,y}(f) = 2\int_{-\infty}^{\infty} \overline{X(t)Y(t+s)} \exp(j\omega s)\, ds;$$

$$S_{y,x}(f) = 2\int_{-\infty}^{\infty} \overline{Y(t)X(t+s)} \exp(j\omega s)\, ds \qquad (5.17)$$

It is now easily shown that

$$S_{y,x}(f) = S^*_{x,y}(f) \qquad (5.17a)$$

when the asterisk denotes the complex conjugate. For

$$\overline{Y(t)X(t+s)} = \overline{Y(t'-s)X(t')} = \overline{X(t')Y(t'-s)} = \overline{X(t)Y(t-s)}$$

Hence

$$S_{y,x}(f) = 2\int_{-\infty}^{\infty} \overline{X(t)Y(t-s)} \exp(j\omega s)\, ds$$

$$= 2\int_{-\infty}^{\infty} \overline{X(t)Y(t+s')} \exp(-j\omega s')\, ds' = S^*_{x,y}(f)$$

as follows from putting $s = -s'$.

5.2 EVALUATION OF $S_x(f)$ OR $S_x(0)$

We here discuss a variety of methods for evaluating $S_x(f)$.

5.2a Evaluation with the Help of the Autocorrelation Function

We show with the help of an example of how the autocorrelation function may sometimes be directly evaluated. In weakly n-type semiconductor material hole–electron pairs are generated either by light or by thermal generation and recombine in a recombination process. As a consequence the number P of holes in the sample will fluctuate around its average value $\bar{P}$. Let $\Delta P = (P - \bar{P})$ be that fluctuation and let the generation–recombination process be characterized by the time constant τ.

We consider an ensemble of identical systems subjected to independent fluctuations. Next we take a subensemble of systems that have $\Delta P = \Delta P_0$ at $t=0$. We then find $\overline{\Delta P}^s$ for this subensemble at a later instant $u(u>0)$ as

$$\overline{\Delta P}^s = \Delta P_0 \exp\left(-\frac{u}{\tau}\right)$$

according to the definition of lifetime. Therefore for $u>0$

$$\Delta P_0 \overline{\Delta P}^s = \Delta P_0^2 \exp\left(-\frac{u}{\tau}\right)$$

$$\overline{\Delta P(0)\Delta P(u)} = \overline{\Delta P_0 \overline{\Delta P}^s} = \overline{\Delta P_0^2}\exp\left(-\frac{u}{\tau}\right) = \overline{\Delta P^2}\exp\left(-\frac{u}{\tau}\right) \quad (5.18)$$

Hence

$$S_p(f) = 4\int_0^\infty \overline{\Delta P^2}\exp\left(-\frac{u}{\tau}\right)\cos\omega u\, du = 4\overline{\Delta P^2}\frac{\tau}{1+\omega^2\tau^2} \quad (5.19)$$

We come back to this problem in Section 5.2f.

5.2b Evaluation of $S_x(0)$ from $\overline{X_\tau^2}$

The most important application of this method is that in a series of random events occurring at the rate $n(t)$

$$S_n(0) = 2\,\mathrm{var}\,n \quad (5.20)$$

where $\mathrm{var}\,n$ is the variance of $n(t)$.

Proof. The number N of events occurring in an interval τ is

$$N = \int_0^\tau n(t)\,dt, \quad \text{or} \quad \bar{N} = \bar{n}\tau$$

since the variations in individual seconds are independent and time averages are equal to ensemble averages. Hence for sufficiently large τ we have for $X_\tau = (\Delta N/\tau)$; since $\overline{\Delta N^2} = \tau\,\mathrm{var}\,n$,

$$\overline{X_\tau^2} = \frac{\overline{\Delta N^2}}{\tau^2} = \frac{\tau\,\mathrm{var}\,n}{\tau^2} = \frac{\mathrm{var}\,n}{\tau}$$

so that $S_n(0) = \lim_{\tau\to\infty} 2\tau\overline{X_\tau^2} = 2\,\mathrm{var}\,n$, as had to be proved.

This powerful theorem has many applications.

Evaluation of $S_x(f)$ or $S_x(0)$

1. Noise in a Saturated Thermionic Diode. We shall prove that the spectral intensity of the fluctuating current $I(t)$ of average value $\bar{I}$ is

$$S_I(0) = 2e\bar{I} \tag{5.21}$$

This is called *Schottky's theorem*.

Proof. Since thermionic emission obeys Poisson statistics, $\operatorname{var} n = \bar{n}$, but $I(t) = en(t)$, where e is the electron charge. Hence

$$\bar{I} = e\bar{n} \,; \quad S_I(0) = e^2 S_n(0) = 2e^2 \operatorname{var} n = 2e^2 \bar{n} = 2e\bar{I}$$

as had to be proved.

2. Noise in a Space-Charge Limited Thermionic Diode. Here the fluctuations are smoothed by the space charge, so that $\operatorname{var} n < \bar{n}$. Writing

$$\operatorname{var} n = \bar{n}\,\Gamma^2 \tag{5.22}$$

where Γ^2 is the space-charge smoothing factor, we have

$$S_I(0) = 2e^2 \operatorname{var} n = 2e^2 \bar{n}\,\Gamma^2 = 2e\bar{I}\,\Gamma^2 \tag{5.23}$$

3. Noise in a Solid-State p^+-n Junction. The device has a characteristic

$$\bar{I} = I_{rs}\left[\exp\left(\frac{eV}{kT}\right) - 1\right] \tag{5.24}$$

It consists of two fluctuating currents $I_{rs}\exp(eV/kT)$ and I_{rs} flowing in opposite directions and fluctuating independently. Since each current satisfies (5.21), we have

$$S_I(0) = 2eI_{rs}\exp\left(\frac{eV}{kT}\right) + 2eI_{rs} = 2e(\bar{I} + 2I_{rs}) \tag{5.25}$$

For $\bar{I} \gg I_{rs}$ this corresponds to (5.21). An alternate way of writing (5.25) is

$$S_I(0) = 2e\bar{I}\,\frac{\exp(eV/kT) + 1}{\exp(eV/kT) - 1} = 2e\bar{I}\coth\left(\frac{eV}{2kT}\right) \tag{5.25a}$$

4. Shot Noise in Transistors. If we neglect small saturation currents, and the emitter, base, and collector currents are I_E, I_B, and I_C, respectively, then

$$S_{I_E}(0) = 2eI_E; \quad S_{I_B}(0) = 2eI_B; \quad S_{I_C}(0) = 2eI_C \tag{5.26}$$

$$S_{I_C, I_B}(0) = 0; \quad S_{I_E, I_C}(0) = 2eI_C \tag{5.26a}$$

Proof. We showed earlier that

$$\operatorname{var} n_C = \overline{n_C}, \quad \operatorname{var} n_B = \overline{n_B}, \quad \overline{\Delta n_C \Delta n_B} = 0$$

so that the fluctuations are independent and Poissonian. Therefore the second and third halves of (5.26) and the first half of (5.26a) are correct. Furthermore

$$I_E = I_B + I_C, \quad \Delta I_E = \Delta I_B + \Delta I_C$$

$$S_{I_E}(0) = S_{I_B}(0) + S_{I_C}(0) + 2S_{I_C, I_B}(0) = 2e(I_B + I_C) = 2eI_E$$

because of the last half of (5.26) and the first half of (5.26a). This proves the first part of (5.26). Moreover

$$\overline{\Delta I_E \Delta I_C} = \overline{\Delta I_B \Delta I_C} + \overline{\Delta I_C^2} = \overline{\Delta I_C^2}$$

and hence

$$S_{I_E, I_C}(0) = S_{I_C}(0) = 2eI_C$$

which proves the second half of (5.26a). This result is important for the common-base transistor circuit.

5.2c Evaluation of $S_x(0)$ from Statistical Considerations

In some cases statistical considerations give enough information to calculate $S_x(0)$. As an example we prove Nyquist's theorem, according to which the spectral intensity $S_V(0)$ of the noise e.m.f. $V(t)$ in series with a resistance R at the temperature T is

$$S_V(0) = 4kTR \tag{5.27}$$

where k is Boltzmann's constant.

Proof. The thermal noise can be described by an r.m.s. e.m.f. $[S_V(0)\Delta f]^{1/2}$ in series with R. If we take the RC circuit of Fig. 5.2a, we see that the frequency interval df gives a contribution

$$\overline{dv_c^2} = S_V(0) df \frac{1/\omega^2 C^2}{R^2 + 1/\omega^2 C^2} = \frac{S_V(0) df}{1 + \omega^2 C^2 R^2}$$

to the mean square value of the voltage v_c across C. Hence

$$\overline{v_c^2} = \int_0^\infty \overline{dv_c^2} = S_V(0) \int_0^\infty \frac{df}{1 + \omega^2 C^2 R^2} = \frac{S_V(0)}{2\pi CR} \int_0^\infty \frac{dx}{1 + x^2} = \frac{S_V(0)}{4CR}$$

Evaluation of $S_x(f)$ or $S_x(0)$

where $x = \omega CR$. However, according to the equipartition law

$$\tfrac{1}{2} C \overline{v_c^2} = \tfrac{1}{2} kT, \quad \text{or} \quad \overline{v_c^2} = \frac{kT}{C} \tag{5.28}$$

Hence $S_V(0) = 4CR\overline{v_c^2} = 4kTR$.

If the noise is represented by an r.m.s. current generator $[S_I(0)\Delta f]^{1/2}$ in parallel with the circuit (Fig. 5.2b) then

$$S_I(0) = \frac{S_V(0)}{R^2} = \frac{4kT}{R} = 4kTg \tag{5.29}$$

where $g = 1/R$.

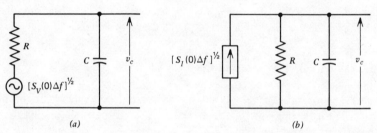

(a)   (b)

Figure 5.2. (a) Thermal noise of an RC circuit, represented by a noise e.m.f. $[S_V(0)\Delta f]^{1/2}$ in series with R. (b) thermal noise e.m.f. of an RC circuit, represented by a noise current generator $[S_I(0)\Delta f]^{1/2}$ in parallel with R.

Equations (5.27) and (5.29) are valid up to infrared frequencies. If hf/kT becomes comparable to unity a quantum correction must be made, according to which

$$S_V(f) = 4R \frac{hf}{\exp(hf/kT) - 1} \tag{5.30}$$

where h is Planck's constant. Usually $(hf/kT) \ll 1$, and then (5.30) reduces to (5.27). The condition $hf_0 = kT$ or $f_0 = (kT/h)$ yields for $T = 300°$ K

$$f_0 = \frac{1.4 \times 10^{-23} \times 300}{6.6 \times 10^{-34}} = 6 \times 10^{12} \text{ Hz}$$

which frequency lies in the infrared. At cryogenic temperatures the frequency f_0 lies at about 100 kMHz.

The reason for the quantum correction is that a tuned circuit tuned at the frequency f can be considered as an harmonic oscillator of frequency f. For low frequencies f the average energy of the harmonic oscillator is kT, but at higher frequencies, where hf is comparable to or larger than kT, the average energy is

$$\overline{E} = \frac{hf}{\exp(hf/kT)-1} = kT\frac{hf/kT}{\exp(hf/kT)-1} \tag{5.31}$$

because the harmonic oscillator can only assume energies $E_v = (v + \tfrac{1}{2})hf$ ($v = 0, 1, 2 \ldots$). The average low-frequency noise energy in the system must thus be multiplied by the factor

$$\frac{hf/kT}{\exp(hf/kT)-1}$$

and hence (5.27) must be multiplied by the same factor. This immediately yields (5.30). For a more elaborate proof see van der Ziel.*

5.2d Evaluation of $S_x(f)$ from Carson's Theorem

Carson's theorem can be stated as follows: Let a stationary random variable $Y(t)$ be the sum of the effects of a large number of independent events, occurring at random instants t_i at the average rate λ, so that

$$Y(t) = \sum_i F(t - t_i) \tag{5.32}$$

where t_i is the instant at which the event starts, so that $F(t - t_i) = 0$ for $t < t_i$ and $F(t - t_i)$ represent the effect of the ith event for $t > t_i$. Defining the Fourier transform $\psi(f)$ of $F(u)$ as

$$\psi(f) = \int_{-\infty}^{\infty} F(u) \exp(-j\omega u) \, du \tag{5.33}$$

where $F(u)$ is assumed to be absolutely integrable, the spectral intensity $S_y(f)$ of $Y(t)$ is

$$S_y(f) = 2\lambda |\psi(f)|^2 \tag{5.34}$$

For an elaborate proof see van der Ziel.*

A simple proof can be given as follows: Since the events occur at random instants, it is sufficient to evaluate the spectral intensity $S_i(f)$ of a *single* event occurring at the instant t_i in the interval $0 \leq t \leq T$. We then

---

*A. van der Ziel, *Noise: Sources, Characterization, Measurements*, Prentice Hall, Englewood Cliffs, N. J., 1970.

Evaluation of $S_x(f)$ or $S_x(0)$

have for large T and $t_i \cong \tfrac{1}{2}T$

$$a_n = \frac{1}{T}\int_0^T F(t-t_i)\exp(-j\omega_n t)\,dt = \frac{\exp(-j\omega_n t_i)}{T}\int_{-t_i}^{T-t_i} F(u)\exp(-j\omega_n u)\,du$$

$$\cong \frac{\exp(-j\omega_n t_i)}{T}\int_{-\infty}^{\infty} F(u)\exp(-j\omega_n u)\,du = \frac{\exp(-j\omega_n t_i)}{T}\psi(f_n)$$

Therefore for a single event

$$S_i(f_n) = 2 a_n a_n^* T = \frac{2|\psi(f_n)|^2}{T}$$

However, on the average there are λT events in the interval $0 \leqslant t \leqslant T$ and hence

$$S_y(f) = \lambda T S_i(f_n) = 2\lambda |\psi(f_n)|^2$$

as had to be proved.

As an example we consider the motion of an electron in a planar saturated thermionic diode at an anode potential V_a. It is easily shown that the transit time τ of an electron flowing from cathode to anode is

$$\tau = \frac{2d}{v_a} = \frac{2d}{(2eV_a/m)^{1/2}} \tag{5.35}$$

where v_a is the velocity of electrons at the anode. This follows directly from the fact that the electrons are uniformly accelerated, so that their average velocity is $v_a/2$.

It is also easily shown that the current $i(t)$ due to an electron emitted at the instant t_i is

$$i(t) = \frac{2e(t-t_i)}{\tau^2} \quad \text{for} \quad t_i \leqslant t \leqslant t_i + \tau \tag{5.36}$$

and zero otherwise. For a proof see van der Ziel.* That (5.36) is correct is seen from the fact that

$$\int_0^\tau i(t)\,dt = e$$

---

*A. van der Ziel, *Noise*, Prentice Hall, Englewood Cliffs, N. J., 1954.

that is, a full electron charge is transferred during the transit time τ. Hence

$$\psi(f) = \int_0^\tau \frac{2eu}{\tau^2} \exp(-j\omega u)\, du = e\phi_3(j\omega\tau) \tag{5.37}$$

where

$$\phi_3(j\omega\tau) = \frac{2}{(j\omega\tau)^2}\left[1 - \exp(-j\omega\tau) - j\omega\tau \exp(-j\omega\tau)\right] \tag{5.37a}$$

Consequently, since $\lambda = (\bar{I}/e)$

$$S_I(f) = \frac{2\bar{I}}{e}|\psi(f)|^2 = 2e\bar{I}|\phi_3(j\omega\tau)|^2 \tag{5.38}$$

Since τ is of the order of 10^{-9} s, and $|\phi_3(j\omega\tau)|^2$ is close to unity for $\omega\tau < \pi$, $S_i(f)$ is white up to about a few hundred MHz.

5.2e Evaluation of $S_x(f)$ by the Langevin Method

In the Langevin method one writes the macroscopic equation of the system under investigation and then puts in the left-hand side a white random-source function describing the noise. One then makes a Fourier analysis of this equation, determines the Fourier coefficients, and from it, the spectrum. Usually enough information is available to evaluate the spectral intensity of the random-source function.

We illustrate this with a few examples:

1. RC Circuit Driven by Thermal Noise (Fig. 5.2a). The equation of the circuit is

$$Ri + \int \frac{i\, dt}{C} = E(t) \tag{5.39}$$

where $E(t)$ is the white-noise e.m.f. in series with R. For $0 \leqslant t \leqslant T$ we substitute a Fourier series

$$E(t) = \sum_{n=-\infty}^{\infty} a_n \exp(j\omega_n t); \qquad i(t) = \sum_{n=-\infty}^{\infty} b_n \exp(j\omega_n t)$$

Substituting into (5.39) and equating terms with $\exp(j\omega_n t)$ yields

$$b_n\left(R + \frac{1}{j\omega_n C}\right) = a_n, \quad \text{or} \quad b_n = a_n \frac{j\omega_n C}{1 + j\omega_n CR}$$

Hence

$$S_I(f) = \frac{S_E(f)\omega_n^2 C^2}{1+\omega_n^2 C^2 R^2} = \frac{4kTR}{R^2 + 1/(\omega_n^2 C^2)} \tag{5.40}$$

since $S_E(f)$ follows from Nyquist's theorem.

2. Spontaneous Temperature Fluctuations of a Small Body. Let a small body have a heat capacity C_H, and a heat conductance g_H to its surroundings, and let θ be the temperature fluctuation. The Langevin equation is now

$$C_H \frac{d\theta}{dt} + g_H \theta = H(t) \tag{5.41}$$

where $H(t)$ is an unknown white random-source function.

We make a Fourier analysis for $0 \leq t \leq T$. Hence

$$H(t) = \sum_{n=-\infty}^{\infty} a_n \exp(j\omega_n t); \quad \theta(t) = \sum_{n=-\infty}^{\infty} b_n \exp(j\omega_n t)$$

Substituting into (5.41) and equating terms with $\exp(j\omega_n t)$ yields

$$b_n(j\omega_n C_H + g_H) = a_n, \quad b_n = \frac{a_n}{j\omega_n C_H + g_H}$$

so that

$$S_\theta(f) = \frac{S_H(f)}{\omega_n^2 C_H^2 + g_H^2} \tag{5.42}$$

We shall prove in a moment that

$$\overline{\theta^2} = \frac{kT^2}{C_H} \tag{5.43}$$

This provides sufficient information to evaluate $S_H(f)$. For since $S_H(f)$ has a white spectrum

$$\overline{\theta^2} = \int_0^\infty S_\theta(f) \, df = S_H(0) \int_0^\infty \frac{df}{\omega^2 C_H^2 + g_H^2} = \frac{S_H(0)}{4g_H C_H}$$

Hence

$$S_H(0) = 4g_H C_H \overline{\theta^2} = 4kT^2 g_H \tag{5.44}$$

and

$$S_\theta(f) = \frac{4kT^2 g_H}{\omega^2 C_H^2 + g_H^2} \tag{5.44a}$$

Next we prove (5.43). We take a large number of identical small bodies and keep them in equilibrium with a temperature bath of temperature T. Let each system be able to assume energies $E_i (i=1,2\ldots)$. Since the probability that a system has an energy E_i is $A\exp(-E_i/kT)$, and this distribution must be normalized, we have

$$\sum_i A \exp\left(-\frac{E_i}{kT}\right) = 1, \quad \text{or} \quad A = \frac{1}{\sum_i \exp(-E_i/kT)}$$

Therefore, the average energy is

$$\overline{E_i} = \sum_i E_i A \exp\left(-\frac{E_i}{kT}\right) = \frac{\sum_i E_i \exp(-E_i/kT)}{\sum_i \exp(-E_i/kT)}$$

The heat capacity is therefore

$$C_H = \frac{\partial \overline{E_i}}{\partial T} = \frac{1}{kT^2} \left\{ \frac{\sum_i E_i^2 \exp(-E_i/kT)}{\sum_i \exp(-E_i/kT)} - \left[\frac{\sum_i E_i \exp(-E_i/kT)}{\sum_i \exp(-E_i/kT)}\right]^2 \right\}$$

$$= \frac{\overline{E_i^2} - (\overline{E_i})^2}{kT^2}, \quad \text{or} \quad \overline{E_i^2} - (\overline{E_i})^2 = C_H kT^2$$

However,

$$E_i - \overline{E_i} = C_H \theta \quad \text{or} \quad \overline{E_i^2} - (\overline{E_i})^2 = C_H^2 \overline{\theta^2} \quad \text{or} \quad \overline{\theta^2} = \frac{kT^2}{C_H}$$

as had to be proved.

5.2f Generation–Recombination Noise

In semiconductors one encounters noise due to the generation and recombination of carriers. We have already seen a particular example in Section

Evaluation of $S_x(f)$ or $S_x(0)$

5.2a. The general equation for the carrier number N in a semiconductor sample is

$$\frac{dN}{dt} = g(N) - r(N) \tag{5.45}$$

where $g(N)$ and $r(N)$, describing the generation and the recombination rates of carriers, respectively, are known functions of N. The equilibrium value N_0 of N follows from the condition $(dN/dt)=0$ for $N=N_0$ or

$$g(N_0) = r(N_0) \tag{5.45a}$$

We now look for small fluctuations of N around the equilibrium value N_0, put $N=(N_0+\Delta N)$ and make a Taylor expansion of $g(N)$ and $r(N)$ around N_0. This yields

$$\frac{d\Delta N}{dt} = -\left(\frac{dr}{dN} - \frac{dg}{dN}\right)\bigg|_{N_0} \Delta N + \Delta g(t) - \Delta r(t) = -\frac{\Delta N}{\tau} + \Delta g(t) - \Delta r(t)$$

$$\tag{5.46}$$

where

$$\frac{1}{\tau} = \left(\frac{dr}{dN} - \frac{dg}{dN}\right)\bigg|_{N_0} \tag{5.46a}$$

τ is the time constant of the system and $\Delta g(t)$ and $\Delta r(t)$ are white random-source functions describing the shot noise in the rates $g[N(t)]$ and $r[N(t)]$.

Making a Fourier expansion for $0 \leq t \leq T$ by putting

$$\Delta N(t) = \sum_{n=-\infty}^{\infty} a_n \exp(j\omega_n t); \quad \Delta g(t) = \sum_{n=-\infty}^{\infty} b_n \exp(j\omega_n t);$$

$$\Delta r(t) = \sum_{n=-\infty}^{\infty} c_n \exp(j\omega_n t)$$

we substitute into (5.46). Equating terms with $\exp(j\omega_n t)$ yields

$$\left(j\omega_n + \frac{1}{\tau}\right)a_n = b_n - c_n; \quad a_n = \frac{\tau}{1+j\omega_n \tau}(b_n - c_n)$$

or

$$S_N(f) = \frac{\tau^2}{1+\omega^2\tau^2}[S_g(f) + S_r(f)] = \frac{4g(N_0)\tau^2}{1+\omega^2\tau^2} \tag{5.47}$$

because $S_g(f) = S_r(f) = 2g(N_0) = 2r(N_0)$, since they represent shot-noise sources.

However, in analogy with (5.19) we may also write

$$S_N(f) = 4\overline{\Delta N^2} \frac{\tau}{1+\omega^2\tau^2} \tag{5.48}$$

Equating this to (5.47) yields

$$\overline{\Delta N^2} = g(N_0)\tau = \frac{g(N_0)}{(dr/dN - dg/dN)|_{N_0}} \tag{5.49}$$

This result can also be proven from the master equation.*

As an example, we consider weakly n-type material. Here

$$g(N) = g_0 = \text{const}; \quad r(N) = \rho NP = \rho N(N - N_d) \tag{5.50}$$

where g_0 and ρ are constants and N_d is the donor concentration. The equilibrium condition is

$$g_0 = \rho N_0 P_0 = \rho N_0 (N_0 - N_d) \tag{5.50a}$$

Furthermore

$$\frac{1}{\tau} = \left(\frac{dr}{dN} - \frac{dg}{dN}\right)_{N_0} = \rho(2N_0 - N_d) = \rho(N_0 + P_0) \tag{5.50b}$$

and hence

$$\overline{\Delta N^2} = g_0\tau = \frac{g_0}{\rho(N_0 + P_0)} = \frac{N_0 P_0}{N_0 + P_0} \tag{5.50c}$$

These fluctuations in number can be detected by passing a current I_0 through the sample or by applying a voltage V to the sample. If the sample contains only electrons of density n, then the conductance

$$g = e\mu_n n \frac{A}{L} = \frac{e\mu_n N}{L^2}$$

where $N = nAL$ is the total number of carriers, A the cross section of the sample, L the sample length, and μ_n is the electron mobility. Hence

$$I_0 = gV = \frac{e\mu_n N_0 V}{L^2}; \quad \Delta I = \frac{e\mu_n V}{L^2} \Delta N = \frac{I_0}{N_0} \Delta N$$

*Compare, for example, with A. van der Ziel, *Noise: Sources, Characterization, Measurements*, Prentice Hall, Englewood Cliffs, N. J., 1970.

Evaluation of $S_x(f)$ or $S_x(0)$

so that

$$S_I(f) = \left(\frac{e\mu_n V}{L^2}\right)^2 S_N(f) = 4\left(\frac{e\mu_n V}{L^2}\right)^2 \overline{\Delta N^2} \frac{\tau}{1+\omega^2\tau^2} \qquad (5.51)$$

or

$$S_I(f) = \frac{I_0^2}{N_0^2} S_N(f) = 4\frac{I_0^2}{N_0^2} \overline{\Delta N^2} \frac{\tau}{1+\omega^2\tau^2} \qquad (5.51a)$$

If both electrons and holes are present in numbers N_0 and P_0 and fluctuations ΔN and ΔP, then

$$I = e(\mu_n N_0 + \mu_p P_0)\frac{V}{L^2} = \left(\frac{e\mu_n V}{L^2}\right)\left(N_0 + \frac{\mu_p}{\mu_n}P_0\right)$$

$$\Delta I = \left(\frac{e\mu_n V}{L^2}\right)\left(\Delta N + \frac{\mu_p}{\mu_n}\Delta P\right)$$

Therefore, if $S_{N,N}(f)$, $S_{N,P}(f)$, and $S_{P,P}(f)$ are the spectra

$$S_i(f) = \left(\frac{e\mu_n V}{L^2}\right)^2 \left[S_{N,N}(f) + \left(\frac{\mu_p}{\mu_n}\right)^2 S_{P,P}(f) + 2\left(\frac{\mu_p}{\mu_n}\right)\mathrm{Re}\, S_{N,P}(f)\right] \qquad (5.52)$$

where Re stands for "real part of." If $\Delta N = \Delta P$, $S_{N,N} = S_{P,P} = S_{N,P}$, and

$$S_I(f) = 4\left(\frac{e\mu_n V}{L^2}\right)^2 \left(\frac{\mu_n + \mu_p}{\mu_n}\right)^2 \overline{\Delta N^2} \frac{\tau}{1+\omega^2\tau^2} \qquad (5.52a)$$

6

NOISE CHARACTERIZATION DEVICES AND AMPLIFIERS

This chapter discusses noise characterization of passive and active networks in terms of the equivalent noise resistance, equivalent noise conductance, equivalent noise temperature, and the noise figure of four-terminal networks, with applications to FETs and transistors (Section 6.1). Sections 6.2 and 6.3 discuss the noise figure of some simple FET and transistor circuits.

6.1 NOISE CHARACTERIZATION

6.1a Two-Terminal Networks

The noise in any passive or active two-terminal network at the temperature T can be represented by an e.m.f. e_n in series with the impedance $Z=(R+jX)$ of the network (Fig. 6.1a). We now equate

$$\overline{e_n^2} = 4kTR_n \Delta f \qquad (6.1)$$

where Δf is the frequency band for which the e.m.f. e_n is defined and call R_n the equivalent noise *resistance* of the network.

If the circuit is a linear passive circuit that has thermal noise sources at the temperature T only, then

$$R_n = R \qquad (6.1a)$$

If other noise sources are present, R_n may differ from R. The same is true for devices that do not have thermal noise or that show thermal noise but are nonlinear.

Noise Characterization

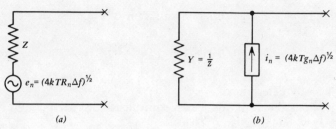

Figure 6.1. (a) Noise of a passive or active impedance Z, represented by an e.m.f. $(4kTR_n\Delta f)^{1/2}$ in series with Z; (b) noise of a passive or active admittance $Y=1/Z$, represented by a current generator $(4kTg_n\Delta f)^{1/2}$ in parallel with Y.

The noise in any passive or active two-terminal network at the temperature T can also be represented by a current generator i_n in parallel with the admittance $Y=1/Z=(g+jb)$ (Fig. 6.1b). We now equate

$$\overline{i_n^2} = 4kTg_n\Delta f \tag{6.2}$$

and call g_n the *equivalent noise conductance* of the device.

If the circuit is a linear passive circuit, and all noise sources are thermal noise sources at the temperature T, then

$$g_n = g \tag{6.2a}$$

If other noise sources are present, or if the system is nonlinear, g_n may differ from g.

It is also possible to equate

$$\overline{i_n^2} = 2eI_{eq}\Delta f \tag{6.3}$$

and call I_{eq} the *equivalent saturated diode* current of the network, This is especially appropriate if all noise sources are shot-noise sources. For example, in a saturated thermionic diode carrying a current I_d

$$I_{eq} = I_d \tag{6.3a}$$

In a space-charge-limited thermionic diode the current fluctuations are smoothed by the space charge. We expressed this by equating

$$\overline{i_n^2} = 2e\Gamma^2 I_d \Delta f$$

where Γ^2 is the space-charge suppression factor of the noise. Hence

$$I_{eq} = \Gamma^2 I_d, \quad \text{or} \quad \Gamma^2 = \frac{I_{eq}}{I_d} \tag{6.3b}$$

so that Γ^2 is easily determined.

Finally, it is possible to introduce an equivalent-noise temperature T_n in (6.1) and (6.2) by equating

$$\overline{e_n^2} = 4kT_n R \Delta f \quad \text{or} \quad T_n = \frac{R_n}{R} T \tag{6.4}$$

$$\overline{i_n^2} = 4kT_n g \Delta f \quad \text{or} \quad T_n = \frac{g_n}{g} T \tag{6.5}$$

This is useful in devices where the current carriers have an equivalent temperature different from the environment, and is exemplified in the case of hot-electron noise in solids or hot-electron noise in a gaseous plasma.

Since one may not know in advance which representation lends itself best to theoretical interpretation, one must be familiar with the conversion from one representation to another. For this and for the measurement of R_n, g_n, I_{eq} or T_n compare with van der Ziel.*

6.1b Four-Terminal Networks

Four-terminal devices like transistors and FETs generally have noise at both the output and the input. The noise must then be represented by an output-current generator i_0 and an input-current generator i_i that are partly correlated (Fig. 6.2a).

Let the input admittance of the device for short-circuited output be denoted by Y_i and the transfer admittance for short-circuited output by Y_m, then the equivalent circuit of the device can be represented by Fig. 6.2b, where the noise e.m.f. $e_n = (i_e/Y_m)$. It is now common practice to equate

$$\overline{e_n^2} = \frac{\overline{i_0^2}}{|Y_m|^2} = 4kTR_n \Delta f \tag{6.6}$$

and call R_n the *equivalent noise resistance* of the device.

---

*A. van der Ziel, *Noise: Sources, Characterization, Measurements,* Prentice Hall, Englewood Cliffs, N. J., 1970, Chapters 3–4.

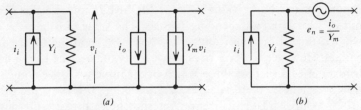

Figure 6.2. (a) Four-terminal active network in which the noise is represented by a current generator i_i in parallel to the input admittance Y_i and a current generator i_0 in parallel to the output, whereas the signal transfer properties are represented by a current generator $Y_m v_i$; (b) same network but the current generator i_0 is replaced by an e.m.f. $e_n = (i_0/Y_m)$ in series with the input.

One can also equate

$$\overline{i_i^2} = 4kTg_{ni}\Delta f \tag{6.7}$$

and call g_{ni} the *equivalent input noise conductance* of the device.

Sometimes the correlation between i_0 and i_i must be taken into account.

We now apply this to FETs operating under saturated conditions. In the junction field effect transistor (JFET) the low-frequency noise is generation–recombination noise and the noise resistance is of the form

$$R_n = \frac{A}{1+\omega^2\tau^2} \tag{6.8}$$

Typically, at room temperature A is of the order of $10^4\ \Omega$ for a good JFET and τ is of the order of 10^{-3} s. In the metal-oxide-semiconductor field effect transistor (MOSFET) the low-frequency noise is flicker noise and R_n is of the form

$$R_n = \frac{B}{f} \tag{6.8a}$$

Typically B is of the order of $10^8\ \Omega$ Hz but it can be much larger in poorer units.

At higher frequencies the noise is thermal noise of the conducting channel. For saturated devices a calculation shows that

$$\overline{i_0^2} = \overline{i_d^2} = \alpha \cdot 4kTg_m\Delta f \tag{6.9}$$

where g_m is the transconductance at saturation. Substituting (6.9) into (6.6) and putting $Y_m = g_m$ we have

$$R_n = \frac{\alpha}{g_m} \tag{6.9a}$$

The parameter α has a value 2/3 for MOSFETs and a value between 1/2 and 2/3 for JFETs, depending on bias. The value of α must be raised somewhat if there is a significant series resistance r_s on the source side of the channel. However, α never rises above unity, unless one has to do with hot-electron effects such as those which occur in very short channels.

At low frequencies there is a gate current I_g in JFETs of the order of 10^{-9}–10^{-12} Amp, and this current gives shot noise because the device is operated as a back-biased diode. It is now not very convenient to use a noise conductance g_{ni}; rather it is better to write

$$\overline{i_i^2} = \overline{i_g^2} = 2eI_g \Delta f \tag{6.10}$$

The only exception is when the gate is floating. There are then two equal and opposite currents I_{g0} flowing across the gate junction and hence

$$\overline{i_i^2} = 4eI_{g0} \Delta f \tag{6.10a}$$

Since the gate of the FET now acts as a floating diode, the input conductance is

$$g_i = \frac{1}{R_g} = \frac{eI_{g0}}{kT} \tag{6.10b}$$

and hence

$$\overline{i_g^2} = 4kTg_i \Delta f \quad \text{or} \quad g_{ni} = g_i \tag{6.10c}$$

For MOSFETs the leakage current is practically zero, so that the problem does not exist.

At high frequencies a gate noise exists both in the JFET and the MOSFET due to the capacitive coupling between channel and gate. A calculation shows that for saturation

$$\overline{i_i^2} = \overline{i_g^2} = \beta \cdot 4kTg_{gs} \Delta f \tag{6.11}$$

where β lies between 1.0 and 4/3 for a JFET, depending on bias, and $\beta = 4/3$ for a MOSFET. Here g_{gs} is the input conductance of the device for short-circuited output; hence $g_{ni} = \beta g_{gs}$. A further calculation shows that for MOSFETs in saturation

$$g_{gs} = \frac{1}{5} \frac{\omega^2 C_{gs}^2}{g_m} \tag{6.11a}$$

where C_{gs} is the gate-source capacitance of the device.

Noise Characterization

For some noise calculations one must take into account that i_g and i_d are somewhat correlated. A calculation shows that the correlation coefficient

$$c = \frac{\overline{i_g i_d^*}}{\left(\overline{i_g^2} \cdot \overline{i_d^2}\right)^{1/2}} \tag{6.12}$$

is imaginary, except at the highest frequencies, and that $|c| \simeq 0.40$ (more exactly, $|c| = 0.395$ for a MOSFET).

We have given here most of the formulas without proof. For details and further references see van der Ziel.* The equivalent circuit is shown in Fig. 6.3.

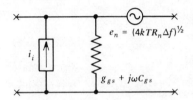

Figure 6.3. High-frequency equivalent noise circuit of a JFET or MOSFET in which the input noise is represented by a current generator i_i in parallel with the input admittance $Y_{gs} = g_{gs} + j\omega C_{gs}$ and the output noise is represented by an e.m.f. $e_n = (4kTR_n \Delta f)^{1/2}$ in series with the input.

We now turn to the transistor in common emitter connection (base as input and collector as output). At low frequencies there is some flicker noise with a $1/f$ spectrum; it can be represented by a current generator $i_i = i_{bf}$ between base and emitter where

$$\overline{i_i^2} = \overline{i_{bf}^2} = \frac{BI_B^\gamma}{f} \Delta f \tag{6.13}$$

Here B is a constant, I_B is the base current, and γ is of the order of unity. This problem is discussed in Chapter 7.

At higher frequencies the noise is shot noise. If I_C is the collector current, then $g_m = (eI_C/kT)$ is the transconductance. Moreover we saw that

$$\overline{i_0^2} = \overline{i_c^2} = 2eI_C \Delta f = 2kTg_m \Delta f \tag{6.14}$$

---

*A. van der Ziel, *Noise: Sources, Characterization, Measurements*, Prentice Hall, Englewood Cliffs, N. J. 1970.

was the collector noise. Hence

$$\overline{e_n^2} = \frac{\overline{i_c^2}}{g_m^2} = \frac{2kT\Delta f}{g_m} \quad (6.14a)$$

so that the noise resistance R_n of the device is

$$R_n = \frac{1}{2g_m} \quad (6.14b)$$

The base current I_B shows full-shot noise. The input conductance g_{be} for short-circuited output is equal to eI_B/kT and hence

$$\overline{i_i^2} = \overline{i_b^2} = 2eI_B\Delta f = 2kTg_{be}\Delta f \quad (6.15)$$

Consequently, the input-noise conductance is

$$g_{ni} = g_{be}/2 \quad (6.15a)$$

The equivalent circuit is shown in Fig. 6.4. To give an approximate high-frequency representation, the input capacitance C_{be} has been added. For a more detailed discussion see van der Ziel.*

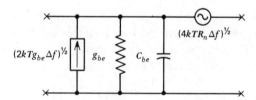

Figure 6.4. High-frequency equivalent noise circuit of a transistor in which the input noise is represented by a current generator $(2kTg_{be}\Delta f)^{1/2}$ in parallel with the input admittance $g_{be} + j\omega C_{be}$ and the output noise is represented by an e.m.f. $(4kTR_n\Delta f)^{1/2}$ in series with the input.

6.1c Noise Figure

The noise figure F of an amplifier is defined as

$$F = \frac{\text{Noise output power of amplifier}}{\text{Part due to the source resistance } R_s} \quad (6.16)$$

---

*A. van der Ziel, *Noise: Sources, Characterization, Measurements*, Prentice Hall, Englewood Cliffs, N. J., 1970.

Noise Characterization

We now refer all noise sources back to the input (Figs. 6.5a, b) and represent them by a current generator $(\overline{i_{eq}^2})^{1/2}$ in parallel with g_s or by a noise e.m.f. $(\overline{e_{eq}^2})^{1/2}$ in series with R_s. Then

$$F = \frac{\overline{i_{eq}^2}}{4kT\Delta f g_s} = \frac{\overline{e_{eq}^2}}{4kTR_s\Delta f} \tag{6.16a}$$

This suggests the following simple method for measuring noise figure. A saturated thermionic diode D is connected in parallel with R_s and so much current I_d is passed through D that the noise-output power of the amplifier is doubled. Then if $X_s = 0$,

$$\overline{i_{eq}^2} = 2eI_d\Delta f$$

and hence

$$F = \frac{e}{2kT}I_d R_s \tag{6.17}$$

We now calculate the noise figure of an FET amplifier stage at frequencies such that the gate noise is negligible. We thus have the circuit of Fig. 6.6a. Here R_s is the source impedance, which has thermal noise, and the device noise is characterized by the noise resistance R_n. Hence

$$F = \frac{4kTR_s\Delta f + 4kTR_n\Delta f}{4kTR_s\Delta f} = 1 + \frac{R_n}{R_s} \tag{6.18}$$

In practice one wants F to be as close to unity as possible. This means

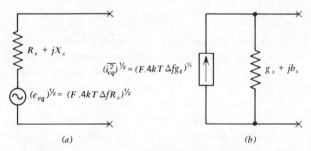

Figure 6.5. Definition of noise figure F: (a) amplifier noise represented by an e.m.f. $(\overline{e_{eq}^2})^{1/2} = (F \cdot 4kT\Delta fR_s)^{1/2}$ in series with the source impedance $Z_s = R_s + jX_s$; (b) amplifier noise represented by a current generator $(\overline{i_{eq}^2})^{1/2} = (F \cdot 4kT\Delta fg_s)^{1/2}$ in parallel to the source admittance $Y_s = (g_s + jb_s)$.

$R_s \gg R_n$. In some cases this cannot be done, because the input circuit may have to satisfy bandwidth restrictions.

In the JFET there is another reason why R_s cannot be made too large and that is that the effect of the gate leakage current I_g becomes important. We illustrate this in Fig. 6.6b. Here R_g is defined by the relation $1/R_g = (dI_g/dV_g)$. For large values fo R_s and R_g, of the order of 10^8–10^{12} Ω, the e.m.f. $(4kTR_n\Delta f)^{1/2}$ gives a negligible contribution to the noise figure and therefore it can be ignored. The noise figure F may therefore be written

$$F = \frac{4kT\Delta f/R_s + 2eI_g\Delta f}{4kT\Delta f/R_s} = 1 + \frac{e}{2kT}I_gR_s \cong 1 + 20I_gR_s \quad (6.19)$$

For $I_g = 10^{-10}$ A and $R_s = 10^{10}$ Ω, $F = 21$. This means that R_s should be chosen smaller than 10^8 Ω in this case if one wants to make F close to unity.

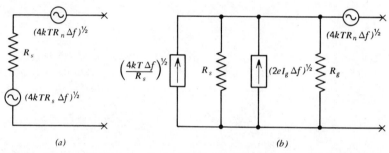

Figure 6.6. Noise-equivalent circuit of a JFET: (a) at low-impedance levels the device noise is represented by an e.m.f. $(4kTR_n\Delta f)^{1/2}$ in series with the input; (b) at high-impedance levels the device noise is represented by a current generator $(2eI_g\Delta f)^{1/2}$ in parallel to the differential input resistance R_g and the e.m.f. $(4kTR_n\Delta f)^{1/2}$ in series with the input.

In cascaded amplifiers the noise of subsequent stages must also be taken into account. Often the most important noise contribution is the contribution of the noise of the load resistance R_L in the output of the first stage. We now wish to point out that its effect is easily included with the help of the circuit of Fig. 6.7. Here the noise at the output of the device is represented by a current generator $(4kTR_n\Delta f g_m^2)^{1/2}$ and the noise of R_L by a current generator $(4kT\Delta f/R_L)^{1/2}$. We now define a new noise resistance R_n' by

$$4kTR_n'\Delta f g_m^2 = 4kTR_n\Delta f g_m^2 + \frac{4kT\Delta f}{R_L} \quad (6.20)$$

Noise Characterization

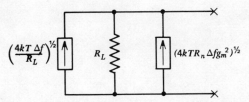

Figure 6.7. Evaluation of the effect of the thermal noise of the load resistance R_L in the output of an FET amplifier stage with an FET noise resistance R_n.

so that

$$R'_n = R_n + \frac{1}{g_m^2 R_L} \tag{6.20a}$$

This expression must now be used in the noise-figure calculations; that is, (6.18) must be replaced by

$$F = 1 + \frac{R'_n}{R_s} = 1 + \frac{R_n}{R_s} + \frac{1}{g_m^2 R_s R_L} \tag{6.21}$$

For example if $R_n = 100\ \Omega$, $R_s = 1000\ \Omega$, $R_L = 2000\ \Omega$, $g_m = 5.0$ mmhos, then $F = 1.12$ of which the part 0.02 comes from the last term in (6.21).

6.1d Friiss's Formula

When one wants to find the noise figure of several stages connected one behind the other (cascade connection), one must establish rules where the load resistance R_L of the interstage network must be counted. The general convention is to count the noise of R_L as belonging to the next stage. We followed this rule in the derivation of (6.18).

We must define the available power gain G_{av} of an amplifier stage. We define the available power P_{av_s} of a signal source consisting of an e.m.f. v_s in series with an internal resistance R_s as the power that can be fed into a matched load. This yields

$$P_{av_s} = \frac{1}{8} \frac{|v_s|^2}{R_s} \tag{6.22}$$

The power gain G of an amplifier stage is now defined as

$$G = \frac{\text{Output power fed into load}}{\text{Power available at source}} \tag{6.23}$$

If the load is matched to the output of the amplifier, the power gain is called the *available power gain*

$$G_{av} = \frac{\text{Output power fed into matched load}}{\text{Power available at source}} \quad (6.23a)$$

There is now an expression for the total noise figure F of the cascaded amplifier that holds if the following conditions are met:

1. The load resistance R_L of each interstage network is considered as belonging to the next stage.
2. The output conductance g_0 seen by each interstage network (i.e., seen when looking toward the preceding stage) is positive.
3. The ith stage, for the given coupling to the preceding stage, has a noise figure F_i and an available gain G_{av_i}.

If these three conditions are valid, the noise figure F of the combination is

$$F = 1 + (F_1 - 1) + \frac{F_2 - 1}{G_{av_1}} + \frac{F_3 - 1}{G_{av_1} G_{av_2}} \quad (6.24)$$

This is known as *Friiss's formula*. We do not use it for the following reasons:

1. It is very clumsy to handle when g_0 is near zero.
2. The circuits we discuss are sufficiently simple for direct evaluation.

6.2 APPLICATIONS TO FET CIRCUITS

6.2a Determination of R_n and I_g in FETs

We now discuss methods for determining the noise resistance R_n and for determining the gate current in JFETs. The circuit is shown in Fig. 6.8. Here the gate is connected to ground via a large bias resistance R_g. The gate is connected to a switch S by means of a blocking capacitor C_b of sufficiently low impedance. S has three positions:

POSITION 1. Connected to ground.

POSITION 2. Connected to a suitable chosen resistance R.

POSITION 3. Connected to a suitably chosen capacitor C.

The output of the FET feeds into a high-gain amplifier that is connected to a power meter.

Applications to FET Circuits

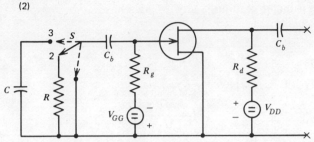

Figure 6.8. Measurement of noise resistance R_n and gate current I_g of a JFET: (a) by comparing the output noise for the switch S in positions 1 and 2 one can evaluate R_n; (b) by comparing the output noise for the switch S in positions 1 and 3 and comparing with the results of the previous measurement one can determine I_g.

We first discuss the measurement of R_n. Here R_g is so chosen that $R_g > 100R$. In position 1 the gate is short-circuited and the power meter reading M_1 corresponds to the noise of R_n, which has a mean square value $4kTR_n\Delta f$. In position 2 the switch is connected to the resistance R, and hence the equivalent noise at the input now has a mean square value $4kT(R+R_n)\Delta f$. If the power meter reading is now M_2, we have

$$\frac{M_2}{M_1} = \frac{4kT(R+R_n)\Delta f}{4kTR_n\Delta f} = 1 + \frac{R}{R_n}$$

so that

$$R_n = \frac{M_1}{M_2 - M_1} R \quad (6.25)$$

For a good measurement, M_1 and $(M_2 - M_1)$ should be comparable.

When one carries out the measurement, one obtains good agreement with theory at higher frequencies, but at lower frequencies the noise increases with decreasing frequency due to flicker noise in MOSFETs or generation–recombination noise in JFETs.

To measure the gate current I_g we make the resistance R_g much larger, for example, $R_g = 10^{10} - 10^{12}$ Ω. The gate current I_g gives a gate noise-current generator $(2eI_g\Delta f)^{1/2}$ in parallel to the input of the FET. If we now turn the switch in position 3, then the noise developed across C has a mean square value $\overline{v^2}$

$$\overline{v^2} = \frac{2eI_g\Delta f}{\omega^2 C^2} = 4kTR'_n\Delta f; \quad R'_n = \frac{e}{2kT}\frac{I_g}{\omega^2 C^2} \quad (6.26)$$

If the output meter reading is now M_3, we have

$$\frac{R_n' + R_n}{R_n} = \frac{M_3}{M_1}; \quad \frac{R_n'}{R_n} = \frac{M_3 - M_1}{M_1} \quad (6.26a)$$

Substituting for R_n we have

$$R_n' = \frac{M_3 - M_1}{M_2 - M_1} R \quad (6.26b)$$

It is now most suitable if M_3 and M_2 are comparable, and M_1 is small in comparison with M_3 and M_2. In that case

$$I_g = \frac{2kT}{e} \omega^2 C^2 R_n' \simeq \frac{2kT}{e} \omega^2 C^2 \frac{M_3}{M_2} R \quad (6.26c)$$

Several conditions have to be met in order to keep the method simple:

1. The noise of $1/R$ must be large in comparison with the noise of I_g, or

$$\frac{4kT}{R} \gg 2eI_g \quad (6.27)$$

2. The noise of $1/R_g$ must be small in comparison with the noise of I_g, or

$$\frac{4kT}{R_g} \ll 2eI_g \quad (6.27a)$$

3. C must be so chosen that R_n' and R are comparable.

Example. $I_g = 10^{-11}$ A. Then the first term in (6.27) is 100 times as large as the second term if $R \simeq 5 \times 10^7 \, \Omega$, and the first term in (6.27a) is 100 times as small as the second term if $R_g \simeq 5 \times 10^{11} \, \Omega$. If the frequency of measurement is 10 Hz, then $C = 30$ pF. If R is chosen smaller, C can be made larger.

6.2b The Source Follower and the Common-Gate FET

We first discuss the source follower. The circuit is shown in Fig. 6.9a. The load resistance R_L is split into two parts, R_{L1} and R_{L2}, to provide proper gate bias; the gate resistor R_g supplies that voltage to the gate. Figure 6.9b shows the equivalent circuit. We see from it that the voltage gain g_v is less than unity

$$g_v = \frac{v_s}{v_g} = \frac{R_L}{1/g_m + R_L} = \frac{g_m R_L}{1 + g_m R_L} \quad (6.28)$$

Here $R_L = (R_{L1} + R_{L2})$.

Applications to FET Circuits

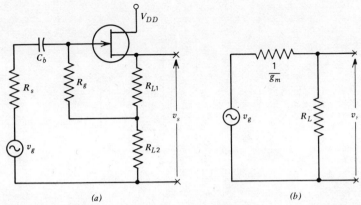

Figure 6.9. (a) Circuit arrangement for a source follower; (b) equivalent circuit for the signal.

If the noise of R_g is neglected and R_n is the noise resistance of the FET, the equivalent circuit for the output circuit itself is shown in Fig. 6.7. Consequently, the noise resistance R'_n of the complete circuit is as given by (6.20a)

$$R'_n = R_n + \frac{1}{g_m^2 R_L} \qquad (6.29)$$

and the noise figure of the complete circuit is

$$F = 1 + \frac{R'_n}{R_s} = 1 + \frac{R_n}{R_s} + \frac{1}{g_m^2 R_s R_L} \qquad (6.30)$$

just as in (6.21).

Although the source follower has a voltage gain slightly less than unity, it serves the useful function of transforming from a high-impedance to low-impedance level with little loss of signal.

Next we turn to the common-gate circuit (Fig. 6.10a). It has the source as input electrode and the drain as output electrode. The signal source looks into an input impedance $Z_{in} = 1/g_m$, and the signal-transfer properties can be characterized by a current generator $g_m v_s$, where v_s is the a.c. voltage at the source electrode.

To simplify the discussion, we short-circuit the output; this is allowed since the noise of the load resistance R_L is counted as belonging to the next stage. First we consider the device noise i alone. It gives an input voltage

$$v_s = -i \frac{R_s \cdot 1/g_m}{R_s + 1/g_m} = -\frac{iR_s}{1 + g_m R_s} \qquad (6.31)$$

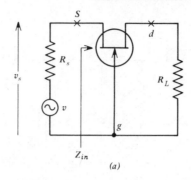

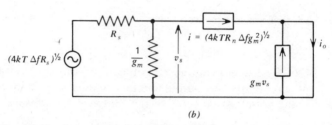

Figure 6.10. (a) Circuit arrangement for the common-gate circuit; (b) equivalent circuit of the common-gate circuit.

and hence a short-circuited output current

$$i'_0 = i + g_m v_s = i\left(1 - \frac{g_m R_s}{1 + g_m R_s}\right) = \frac{i}{1 + g_m R_s} \quad (6.31a)$$

Next we consider the noise of R_s. It gives an input voltage

$$v'_s = (4kTR_s \Delta f)^{1/2} \frac{1/g_m}{1/g_m + R_s} = \frac{(4kTR_s \Delta f)^{1/2}}{1 + g_m R_s} \quad (6.32)$$

and hence a short-circuited output current $i''_0 = (4kTR_s \Delta f)^{1/2} g_m/(1 + g_m R_s)$. Hence adding both noise currents quadratically,

$$\overline{i_0^2} = (4kTR_s \Delta f + 4kTR_n \Delta f)\frac{g_m^2}{(1 + g_m R_s)^2} \quad (6.33)$$

so that the noise figure is

$$F = 1 + \frac{R_n}{R_s} \quad (6.33a)$$

just as for the common source circuit.

Applications to FET Circuits

We notice that the effect of the current generator i is reduced by a factor $1 + g_m R_s$. This is important for understanding the dual gate FET circuit of Section 6.2d.

6.2c Noise of a Two-Stage FET Amplifier

The circuit is shown in Fig. 6.11a. We assume that $R_{L1} = R_{L2} = R_L$; to simplify matters we consider the output of the second stage short-circuited. We now calculate the noise resistance of $Q_1 + Q_2$. The interstage network plus the FETs can be represented by the equivalent circuit of Fig. 6.11b. We use this circuit to evaluate the noise resistance R'_n of the combination. We see by inspection that

$$\overline{v^2} = 4kTR_{n1}\Delta f g_m^2 R_L^2 + 4kTR_L \Delta f + 4kTR_{n2}\Delta f = 4kTR'_n \Delta f g_m^2 R_L^2 \quad (6.34)$$

according to the definition of R'_n. Hence

$$R'_n = R_{n1} + \frac{R_L + R_{n2}}{g_m^2 R_L^2} \quad (6.34a)$$

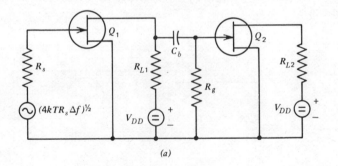

(a)

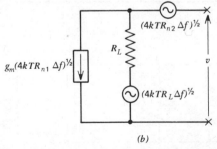

(b)

Figure 6.11. (a) Circuit arrangement for a two-stage FET circuit; (b) noise-equivalent circuit with all noise sources referred to the interstage network.

If $R_{n2} \ll R_L$ this reduces to

$$R'_n \cong R_{n1} + \frac{1}{g_m^2 R_L} \quad (6.34b)$$

so that only the effect of the load resistance R_L is important. The noise figure F is therefore

$$F = 1 + \frac{R'_n}{R_s} = 1 + \frac{1}{R_s}\left(R_{n1} + \frac{1}{g_m^2 R_L} + \frac{R_{n2}}{g_m^2 R_L^2}\right) \quad (6.35)$$

This equation would not have been so easily obtained from Friiss's formula.

6.2d The Dual-Gate FET

The dual-gate FET consists of two FETs Q_1 and Q_2 in a single envelope with the drain d_1 internally connected to the source s_2 so that only the electrodes g_1, s_1, g_2, and d_2 have external leads. For that reason the dual-gate FET is also called the *FET tetrode*. The circuit is shown in Fig. 6.12a. We assume that Q_1 and Q_2 are operated in saturation and that they are identical.

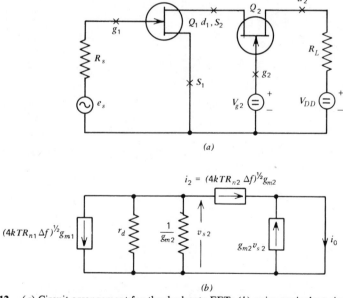

Figure 6.12. (a) Circuit arrangement for the dual-gate FET; (b) noise-equivalent circuit with the noise of the first half of the circuit referred to the output of that FET.

Applications to FET Circuits

We now short-circuit the output and consider the noise of $Q_1 + Q_2$. We introduce the differential resistance r_d of Q_1; it is usually very large but we need it in the calculation. We now have the equivalent circuit of Fig. 6.12b, and use it to calculate the noise resistance of the combination. It is easily seen that

$$\overline{i_0^2} = 4kTR_{n1}\Delta f g_{m1}^2 \left(\frac{r_d}{1+g_{m2}r_d}\right)^2 g_{m2}^2 + \frac{4kTR_{n2}\Delta f g_{m2}^2}{(1+g_{m2}r_d)^2}$$

$$= 4kTR_n'\Delta f g_{m1}^2 \left(\frac{r_d}{1+g_{m2}r_d}\right)^2 g_{m2}^2 \quad (6.36)$$

where $R_{n1} = R_{n2} = R_n$ are the noise resistances and $g_{m1} = g_{m2} = g_m$ the transconductances. Hence

$$R_n' = R_n\left[1 + \frac{1}{g_m^2 r_d^2}\right] \simeq R_n \quad (6.36a)$$

if $g_m^2 r_d^2 \gg 1$, as is usually the case. Therefore, the dual-gate FET has the same noise resistance as the single FET, but it has a much smaller feedback capacitance between the output drain and the input gate than the single FET. Consequently it gives much better stability in high-frequency amplifiers.

6.2e The Pyroelectric Detector with a JFET Amplifier

A pyroelectric detector is a polarized ferroelectric capacitor C with a loss factor $\tan\delta$. If modulated radiation of frequency ω is incident upon the detector, an a.c. voltage of modulation frequency ω is developed across the capacitor (pyroelectric effect). Since the detector is a very high-impedance device, the voltage must be amplified by a high-impedance JFET amplifier. To ensure better stability it is recommended to operate the JFET with a floating base ($I_g = 0$, because two equal but opposite currents I_{g0} flow to the gate).

The pyroelectric detector can thus be used as a radiation detector. Its noise corresponds to thermal noise of its dielectric loss conductance

$$g = \omega C \tan\delta \quad (6.37)$$

The equivalent circuit of detector plus amplifier is shown in Fig. 6.13; here C_i is the input capacitance of the JFET for short-circuited output and $g_{g0} = (eI_{g0}/kT)$ is the input conductance of the JFET with floating gate. We assume that $\omega^2(C+C_i)^2 \gg (g+g_{g0})^2$; this is usually the case, since $(\tan\delta)^2 \ll 1$ and $g_{g0} \ll g$ except perhaps at the lowest frequencies. We then

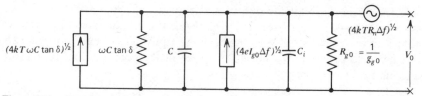

Figure 6.13. Pyroelectric detector having an equivalent circuit consisting of a capacitance C and a conductance $g = \omega C \tan \delta$ in parallel connected to the input of a JFET with floating gate.

have

$$\overline{v_0^2} = \frac{4kT\omega C \tan\delta \Delta f + 4eI_{g0}\Delta f}{\omega^2(C+C_i)^2} + 4kTR_n \Delta f = \frac{4kT\omega C \tan\delta \Delta f}{\omega^2(C+C_i)^2} \cdot F \quad (6.38)$$

where F is the noise figure of the detector. This definition makes sense, since the term

$$\frac{4kT\omega C \tan\delta \Delta f}{\omega^2(C+C_i)^2}$$

comes from the signal source. Hence

$$F = 1 + \frac{e}{kT} \frac{I_{g0}}{\omega C \tan\delta} + R_n \frac{\omega C}{\tan\delta}\left(\frac{C+C_i}{C}\right)^2 \quad (6.38a)$$

If we take $R_n = 10^5\ \Omega$, $\omega = 60/\text{sec}$, $C = 20$ pF, $\tan\delta = 0.01$, $C_i = 5$ pF, $I_{g0} = 10^{-12}$ A, then $F = 4.35$; this value comes mostly from the second term and can be considerably reduced by applying Peltier cooling to the device.

For a MOSFET $I_{g0} \equiv 0$, but R_n is usually much larger because of flicker effect. For example, if

$$R_n = 10^7\ \Omega, \quad \omega = 60/\text{sec}, \quad C = 20\text{ pF}, \quad C_i = 5\text{ pF}, \quad \tan\delta = 0.01,$$

then $F = 2.g$.

In the case of a JFET the capacitance C can be so chosen that F is a minimum. This can be achieved by choosing a pyroelectric detector with the proper dielectric constant or electrode area.

6.2f Field-Effect Transistor Noise at High Frequencies

At high frequencies one must take into account both the gate noise and the drain noise. As a consequence there is now an optimum input tuning and an optimum source conductance for minimum noise figure $F_{\min}$. According to van der Ziel*

$$F_{\min} = 1 + 2R_n g_g + 2\left(1.12 R_n g_g + R_n^2 g_g^2\right)^{1/2} \quad (6.39)$$

Since g_g varies as ω^2 over a wide frequency range and R_n is practically frequency independent in that range, $F_{\min} - 1$ varies as ω at lower frequencies and as ω^2 at higher frequencies. Furthermore his $g_n = \frac{4}{3} g_g (1 - |c|^2) = 1.12 g_g$, since $|c| = 0.395$.

6.3 NOISE IN TRANSISTOR CIRCUITS

6.3a Single Low-Frequency Stage

To find the noise figure at a single low-frequency stage, we modify the equivalent circuit of Fig. 6.4, and assume that the flicker noise is negligible. The modification consists in omitting the capacitance C_{be}, adding the base resistance r_b and its thermal noise and incorporating a source resistance R_s with its thermal noise into the circuit. We then obtain the equivalent circuit of Fig. 6.14a. It, in turn, can be replaced by the equivalent circuit of Fig. 6.14b.

As seen by inspection

$$\overline{v_0^2} = 4kTR_s \Delta f \left(\frac{r_{be}}{R_s + r_b + r_{be}}\right)^2 + 4kTr_b \Delta f \left(\frac{r_{be}}{R_s + r_b + r_{be}}\right)^2$$

$$+ 2kTr_{be} \Delta f \frac{(R_s + r_b)^2}{(R_s + r_b + r_{be})^2} + \frac{2kT\Delta f}{g_m} \quad (6.40)$$

so that the noise figure of the circuit is

$$F = 1 + \frac{r_b}{R_s} + \frac{1}{2} \frac{(R_s + r_b)^2}{R_s r_{be}} + \frac{1}{2g_m R_s} \left(\frac{R_s + r_b + r_{be}}{r_{be}}\right)^2 \quad (6.40a)$$

Usually $r_b \ll R_s$ and $R_s + r_b \ll r_{be}$, so that approximately

$$F \cong 1 + \frac{r_b}{R_s} + \frac{1}{2} \frac{R_s}{r_{be}} + \frac{1}{2g_m R_s} = 1 + \frac{(1 + 2g_m r_b)}{2g_m R_s} + \frac{g_m}{2\beta_F} R_s \quad (6.41)$$

*A. van der Ziel, *Noise: Sources, Characterization, Measurements*, Prentice Hall, Englewood Cliffs, N. J., 1970. This neglects the effects of the correlation conductance g_{cor} and the tuned circuit conductance g_c.

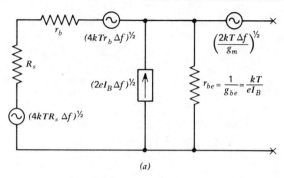

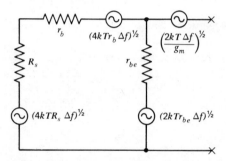

Figure 6.14. (a) Equivalent noise circuit of a transistor in common-emitter connection; (b) alternate equivalent circuit in which the current generator $(2eI_b\Delta f)^{1/2}$ in parallel with r_{be} is replaced by a noise e.m.f. $(2kTr_{be}\Delta f)^{1/2}$ in series with r_{be}.

where $g_m r_{be} = \beta_F$ is the current-amplification factor, so that $r_{be} = (\beta_F/g_m)$. This has a minimum value

$$F_{\min} = 1 + 2\left[\frac{(1+2g_m r_b)}{2g_m} \cdot \frac{g_m}{2\beta_F}\right]^{1/2} = 1 + \left(\frac{1+2g_m r_b}{\beta_F}\right)^{1/2} \quad (6.41a)$$

for

$$R_s = (R_s)_{\text{opt}} = \left[\frac{(1+2g_m r_{be})}{2g_m} \cdot \frac{2\beta_F}{g_m}\right]^{1/2} = \frac{1}{g_m}\left[\beta_F(1+2g_m r_b)\right]^{1/2} \quad (6.41b)$$

An accurate calculation, based on (6.40a), yields

$$F_{\min} = 1 + \frac{1+x}{\beta_F} + \left[\frac{1+2x}{\beta_F} + \left(\frac{1+x}{\beta_F}\right)^2\right]^{1/2} \quad (6.42)$$

Noise in Transistor Circuits

where

$$x = g_m r_b \left(1 + \frac{1}{\beta_F}\right) \tag{6.42a}$$

For large β_F this reduces to (6.41a).

Example. If $g_m = 40$ mmhos, $r_b = 50\ \Omega$, $\beta_F = 100$, find $F_{\min}$ and $(R_s)_{\mathrm{opt}}$ from (6.41a) and (6.41b) and evaluate the difference between (6.42) and (6.41a).

Answer. $F_{\min} = 1 + (0.05)^{1/2} = 1.22$, since $g_m r_b = 2.0$

$$(R_s)_{\mathrm{opt}} = \frac{1}{40 \times 10^{-3}} (500)^{1/2} = 560\ \Omega$$

From (6.42), $F_{\min} = 1.03 + (0.0509)^{1/2} = 1.26$, which is reasonably close.

We now draw the following conclusions:

1. The noise resistance of the circuit for zero source impedance is

$$R_{n0} = \lim_{R_s \to 0} R_s F = r_b + \frac{1}{2g_m} \tag{6.43}$$

according to (6.41). That is, if one wants to measure the noise developed across a low impedance, one should use transistors with a very low base resistance r_b and a relatively large transconductance g_m (i.e., high currents). Noise resistances R_{n0} as low as 50 Ω can be obtained in this manner.

2. The noise can be represented by an input current generator $\overline{(i_{\mathrm{eq}}^2)}^{1/2}$ at the input, where

$$\overline{i_{\mathrm{eq}}^2} = \frac{4kT\Delta f}{R_s} F \tag{6.44}$$

We need this expression for deriving the noise figure of a two-stage transistor amplifier.

3. We can use the argument based on Fig. 6.7 to evaluate the noise resistance R_n' when the effect of the load resistance R_L is taken into account. With $R_L = 0$ the noise resistance is $1/2g_m$, and hence according to (6.29)

$$R_n' = R_n + \frac{1}{g_m^2 R_L} = \frac{1}{2g_m}\left(1 + \frac{2}{g_m R_L}\right) \tag{6.45}$$

Therefore, in (6.40) we must replace g_m by $g_m/[1 + 2/(g_m R_L)]$, and hence in (6.41)

$$F \simeq 1 + \frac{1 + 2/(g_m R_L) + 2g_m r_b}{2g_m R_s} + \frac{g_m R_s}{2\beta_F} \tag{6.45a}$$

so that

$$F_{min} = 1 + \left[\frac{1 + 2/(g_m R_L) + 2g_m r_b}{\beta_F} \right]^{1/2} \quad (6.45b)$$

Often the effect of the load resistance R_L is quite small and no great mistake is made if the term $2/(g_m R_L)$ is omitted.

6.3b Two-Stage Low-Frequency Transistor Amplifier

The circuit is shown in Fig. 6.15; for the sake of simplicity the output of the second stage is short-circuited and the bias circuits are omitted.

According to the previous section the noise of the second stage is given by an equivalent current generator $[F_2(4kT\Delta f)/R_L]^{1/2}$ where

$$F_2 \cong 1 + \frac{R_L}{2r_{be}} + \frac{1 + 2g_m r_b}{2g_m R_L} \quad (6.46)$$

Therefore, in (6.40) we must multiply the last term by the factor

$$1 + 2F_2/(g_m R_L) \quad (6.46a)$$

and hence in (6.41)

$$F = 1 + \frac{1 + 2F_2/(g_m R_L) + 2g_m r_b}{2g_m R_s} + \frac{g_m R_s}{2\beta_F} \quad (6.47)$$

so that

$$F_{min} = 1 + \left[\frac{1 + 2F_2/(g_m R_L) + 2g_m r_b}{\beta_F} \right]^{1/2} \quad (6.47a)$$

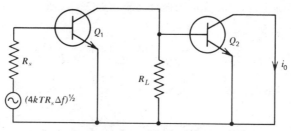

Figure 6.15. Circuit arrangement for a two-stage transistor amplifier.

Noise in Transistor Circuits

Usually the effect of the second stage is quite small and no great error is made in omitting the term altogether.

We have here omitted the finite output resistance $R_0 (1/R_0 = \partial I_C / \partial V_{CE})$ of the transistor. This is allowed if $R_0 \gg R_L$; otherwise a simple correction must be made.

6.3c Emitter-follower Circuit

Figure 6.16 shows an emitter-follower circuit. It has a voltage gain of about unity but a reasonably large power gain. In the figure the bias circuits are omitted.

To calculate the noise figure of a feedback circuit like the emitter follower, we short-circuit the output and count the noise of R_L as belonging to the next stage. The short-circuiting is allowed, for the ratio

$$F = \frac{\text{Voltage noise squared}}{\text{Part due to } R_s} = \frac{\text{current noise squared}}{\text{part due to } R_s}$$

If we now refer the drain noise back to the input as in Fig. 6.14b, we see that the equivalent circuits are the same. Hence the noise figures are also the same.

The incorporation of the noise of the load resistance R_L into the noise figure goes in the same way as in the common emitter circuit and the result is the same.

At high frequencies R_{be} is shunted by the capacitance C_{be} of the transistor. A calculation shows that this results in a large decrease in the power gain G with increasing frequency. As a consequence the circuit is no longer useful.

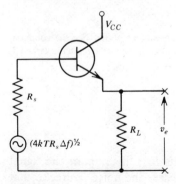

Figure 6.16. Circuit arrangement for the emitter follower.

6.3d Noise in the Common-base Circuit

It is easily seen that the common base and the common emitter circuit have identical noise figures. To that end the output is open-circuited and the equivalent circuit is slightly modified. The modification consists in describing the signal-transfer properties by a current-amplification factor α and a current generator αi_e, where i_e is the current flowing through the emitter junction impedance Z_e; in addition the impedance Z_c of the collector junction is introduced (Fig. 6.17).

The difference between the common emitter and the common base circuits is that the first is grounded at A and the second at B. As long as the noise developed across Z_c is large in comparison with the noise developed in other parts of the circuit, there is no difference in noise figures. Hence the noise figure is again given by (6.40a) and the discussion of Section 6.3a can be taken over directly.

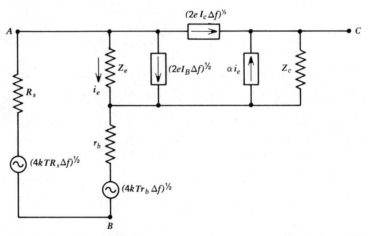

Figure 6.17. Equivalent circuit of the common-base transistor amplifier. For the common-base circuit the circuit is grounded at B, and for the common emitter circuit the circuit is grounded at A, but neither grounding alters the noise figure F.

6.3e Noise at High Frequencies

At high frequencies the capacitance C_{be} shunts R_{be}; this impairs the signal transfer from input to output and hence increases the noise figure. In that

Noise in Transistor Circuits

case one obtains in reasonable approximation*

$$F_{min} = 1 + x + (2x + x^2)^{1/2}, \text{ where } x = g_m r_b \left(\frac{1}{\beta_F} + \frac{f^2}{f_T^2} \right) \quad (6.48)$$

and f_T is the frequency at which $|\beta| = 1$.

6.3f The Darlington Circuit

The transistor is essentially a low-impedance device. However by using a two-stage or three-stage emitter follower one can make a circuit with a high input impedance. The reason is that the emitter-follower circuit steps up the load resistance R_E in the emitter lead by a factor $1 + \beta_F$. Hence in the two-stage emitter follower of Fig. 6.18 the input impedance is

$$Z_{in} = (1 + \beta_{F1})(1 + \beta_{F2})R_E \quad (6.49)$$

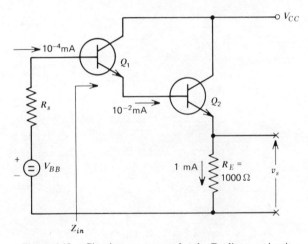

Figure 6.18. Circuit arrangement for the Darlington circuit.

and it has a voltage gain of unity. This circuit is known as the *Darlington circuit*. For $\beta_{F1} = \beta_{F2} = 100$ and $R_E = 1000$ Ω, $Z_{in} \cong 10$ MΩ. In comparison the input resistance of the transistor Q_1 itself is only $r_{be1} = (kT/eI_{B1}) = 260,000$ Ω for $I_{E2} = 1$ mA.

---

*See for example, A. van der Ziel, J. A. Cruz-Emeric, R. D. Livingstone, J. C. Malpass, and D. A. McNamara, *Solid State Electronics*, **19**, 149 (1976), for a discussion and further references. A somewhat more accurate expression is obtained by replacing $g_m r_b$ by $G_1(f) + g_m r_b$, where the function $G_1(f)$ is tabulated by the authors and defined as $G_1(f) = (R_e + R_{cor})/R_{e0}$. Here R_e is the high-frequency emitter resistance, $R_{e0} = kT/eI_E$ is its low-frequency value, and the correlation resistance, R_{cor}, is defined in the paper.

If we use a high source impedance R_s, we may thus simplify the noise calculation since in that case $R_s \gg r_{be1}$ and $R_s \gg r_{b1}$. Furthermore, most of the noise comes from the first stage; the noise of the second stage is negligible. Equation (6.40a) then becomes

$$F = 1 + \frac{1}{2}\frac{R_s}{r_{be1}} + \frac{1}{2g_m R_s}\left(\frac{R_s}{r_{be1}}\right)^2 = 1 + \frac{1}{2}\frac{R_s}{r_{be1}}(1 + 1/\beta_F) \cong 1 + \frac{1}{2}\frac{R_s}{r_{be1}} \quad (6.50)$$

For $I_{E2} = 1$ mA, $r_{be1} = 260,000\ \Omega$; hence for $R_s = 10^7\ \Omega$, $F \cong 20$. The Darlington circuit thus has a large noise figure when operated from a high source impedance. This should be taken into account in applications.

7

FLICKER NOISE AND GENERATION–RECOMBINATION NOISE

Although the exact cause of flicker noise is not known for *all* devices showing it, it is usually assumed that flicker noise in semiconductor resistors, MOSFETs, and transistors is due to generation–recombination noise with a distribution of time constants. Such a distribution is attributed to the interaction of current carriers with traps in the surface oxide. We discuss this explanation in Section 7.1, in Section 7.2 we apply it to MOSFETs, and in Section 7.3, to transistors. In Section 7.4 we discuss flicker noise in carbon resistors.

In Section 7.5 we discuss generation–recombination noise in JFETs. We shall see that there are two causes for it: (a) noise due to generation–recombination centers in the space-charge region of the junctions (the predominant effect at or near room temperature), and (b) noise due to traps or impurity centers in the conducting channel; this effect predominates at low temperatures.

7.1 DERIVATION OF FLICKER NOISE FORMULAS FROM GENERATION–RECOMBINATION NOISE

If a voltage V is applied to a semiconductor sample in which the number N of carriers fluctuates, and the d.c. current flowing through the sample is I_0, then the spectral intensity of the fluctuating current is given by (5.51a) as

$$S_I(f) = 4 \frac{I_0^2}{N_0^2} \overline{\Delta N^2} \frac{\tau}{1 + \omega^2 \tau^2} \qquad (7.1)$$

Here N_0 is the equilibrium number of carriers, $\Delta N = (N - N_0)$, $\overline{\Delta N^2}$ its mean square value and τ the carrier lifetime. This is the well-known generation–recombination noise formula discussed earlier.

In bulk material one can substitute

$$\overline{\Delta N^2} = \beta N_0 \qquad (7.1a)$$

where β is a constant that is not strongly dependent on N_0 but dependent on the carrier statistics of the sample, we obtain

$$S_I(f) = \frac{4\beta I_0^2}{N_0} \cdot \frac{\tau}{1+\omega^2\tau^2} \qquad (7.2)$$

This equation is generally valid; it holds regardless of how the carrier density fluctuation is maintained, but β depends on the process involved.

We now apply this to the case where the electrons in a semiconductor sample interact with traps in the surface oxide layer by tunneling from the surface to the traps in question. In that case it is not so clear that (7.1a) will be valid. For it must be considered that $\overline{\Delta N^2}$ also corresponds to the mean square fluctuation in the number of *trapped* carriers and it is by no means certain that $\overline{\Delta N^2}$ would then be proportional to N_0; in an extreme case it could be independent of N_0 and depend only on the trap density. We can discriminate between these possibilities by determining the dependence of $S_I(f)$ on N_0. If $\overline{\Delta N^2}$ were independent of N_0, $S_I(f)$ would vary as $1/N_0^2$, whereas it varies as $1/N_0$ if (7.1a) is valid.

We assume now that (7.1a) is valid for the interaction of the electrons in a semiconductor with traps in the surface oxide. The constant β is then a measure for the intensity of this interaction. One would expect β to be proportional to the trap density. The trap density itself is proportional to the surface-state density; surface states are centers that can trap electrons but can also generate hole–electron pairs under the proper conditions. β should thus be proportional to the surface-state density.

We shall now show that a $1/f$ spectrum can be obtained with the help of a distribution in time constants τ. The explanation of the time constants τ involves tunneling of carriers through the surface oxide over the distance y between the surface and the oxide trap in question. The value of τ increases very rapidly with increasing y; a calculation shows that it can be expressed as

$$\tau = \tau_0 \exp(\alpha y) \qquad (7.3)$$

where τ_0 is the time constant associated with a surface trap, $\exp(-\alpha y)$ is the tunneling probability, and α is of the order of 10^8 cm^{-1}.

Derivation of Flicker Noise Formulas from Generation–Recombination Noise

To obtain the distribution in time constants τ, we split the surface area into small sections ΔS. Then each section ΔS has its own dominant trap located at a distance y inside the oxide; traps deeper within the oxide have so long a time constant that they can be ignored. Each surface element ΔS thus has its own characteristic lifetime τ. If the dominant traps are distributed at random, then the distribution in y may be written as

$$g(y)\Delta y = \frac{\Delta y}{y_1} \quad \text{for} \quad 0 < y < y_1 \tag{7.4}$$

and zero otherwise. Here y_1 corresponds to the farthest distance between a dominant trap and the surface; it is not infinite, but it is of the order of the average trap distance. Then the corresponding distribution τ is

$$g(\tau)\Delta\tau = \frac{\Delta\tau/\tau}{\ln \tau_1/\tau_0} (\tau_0 < \tau < \tau_1) \tag{7.4a}$$

and zero otherwise, where

$$\tau_1 = \tau_0 \exp(\alpha y_1) \tag{7.4b}$$

We have hereby evaluated the distribution in time constants.

We now average over all surface elements; this corresponds to averaging over all τ values. We thus have instead of (7.2), if $x = \omega\tau$,

$$S_I(f) = 4\frac{\beta I_0^2}{N_0} \int_{\tau_0}^{\tau_1} \frac{\tau}{1+\omega^2\tau^2} \cdot \frac{d\tau/\tau}{\ln \tau_1/\tau_0} = \frac{4\beta I_0^2}{\omega N_0 \ln \tau_1/\tau_0} \int_{\omega\tau_0}^{\omega\tau_1} \frac{dx}{1+x^2}$$

$$= \frac{4\beta I_0^2}{\omega N_0 \ln \tau_1/\tau_0} (\tan^{-1}\omega\tau_1 - \tan^{-1}\omega\tau_0) \tag{7.5}$$

Hence for $1/\tau_1 < \omega < 1/\tau_0$

$$S_I(f) = \frac{\beta I_0^2}{fN_0 \ln \tau_1/\tau_0} = \frac{\alpha I_0^2}{fN_0} \tag{7.5a}$$

where

$$\alpha = \frac{\beta}{\ln \tau_1/\tau_0} \tag{7.5b}$$

The resulting spectrum is thus of the $1/f$ type in that frequency range. It should be seen from (7.5a) that $S_I(f)$ is *inversely proportional* to the number of carriers in the sample. This agrees with experiment.* Any time such agreement is noted, (7.1a) has been verified.

7.2 FLICKER NOISE IN MOSFETS

To evaluate flicker noise in MOSFETs, we start with the Langevin equation for current fluctuations in FETs. It may be written[†]

$$\Delta I = \frac{d}{dx}\left[g(V_0)\Delta V \right] + h(x,t) \tag{7.6}$$

Here $g(V_0)$ is the source conductance per unit length, V_0 the d.c. channel potential at x, $\Delta V(x)$ the noise-voltage distribution along the conducting channel, ΔI the resulting short-circuit noise current, and $h(x,t)$ the distributed noise source present in the sample. If the device is high-frequency short-circuited, $\Delta V = 0$ at $x = 0$ and $x = L$, where L is the length of the conducting channel. Hence by integration

$$\int_0^L \Delta I \, dx = \int_0^L h(x,t) dx, \quad \text{or} \quad \Delta I = \frac{1}{L}\int_0^L h(x,t) dx \tag{7.7}$$

Making a Fourier analysis, we have

$$S_{I_0}(f) = \frac{1}{L^2}\int_0^L \int_0^L S_h(x,x',f) dx \, dx' \tag{7.8}$$

Now fluctuations at x and x' at the instant t are uncorrelated; that is, $S_h(x,x',f)$ is a δ function in $x' - x$, or

$$S_h(x,x',f) = F(x',f)\delta(x' - x) \tag{7.9}$$

Consequently, if we replace x by u

$$S_{I_0}(f) = \frac{1}{L^2}\int_0^L F(u,f) du \tag{7.10}$$

One can often evaluate $S_I(x,f)$ in a section between x and $(x + \Delta x)$ from

---

*F. N. Hooge, *Phys. Lett.*, **29A**, 139 (1969). This author has a different explanation for the noise, however.
[†]A. van der Ziel, *Noise: Sources, Characterization, Measurements*, Prentice Hall, Englewood Cliffs, N. J., 1970.

Flicker Noise in MOSFETs

first principles. If one short-circuits the section Δx, one has

$$S_I(x,f) = \frac{1}{(\Delta x)^2} \int_x^{x+\Delta x} F(u,f)du = \frac{F(x,f)}{\Delta x}, \quad \text{or} \quad F(x,f) = S_I(x,f)\Delta x$$

(7.11)

That is, if we know $S_I(x,f)$ for a section Δx, we can calculate $F(u,f)$ and hence $S_{I_0}(f)$. This is the basis of our flicker and generation–recombination noise calculations.

7.2a Evaluation of the Noise Resistance of a MOSFET

We now apply (7.5a) to a section of the channel with a length Δx and a cross-sectional area $S(x)$. Then $N_0 = n_0(x)S(x)\Delta x$. Hence

$$S_I(x,f) = \frac{\alpha I_0^2}{fN_0} = \frac{\alpha I_0^2}{fn_0(x)S(x)} \cdot \frac{1}{\Delta x}$$

or

$$F(x,f) = \frac{\alpha I_0^2}{fn_0(x)S(x)} = \frac{e\alpha\mu I_0}{f} \frac{dV_0}{dx} \quad (7.12)$$

since $I_0 = e\mu n_0(x)S(x)(dV_0/dx)$. Hence

$$S_{I_0}(f) = \frac{e\mu\bar{\alpha}I_0}{L^2 f} \int_0^L \frac{dV_0}{dx} dx = \frac{e\mu\bar{\alpha}I_0 V_d}{L^2 f} \quad (7.13)$$

We have replaced α by its average value $\bar{\alpha}$; this takes into account the possibility that α is a slow function of x.

Equation (7.13) holds if the drain voltage $V_d < (V_g - V_T)$, where V_g is the gate voltage and V_T the turn-on voltage of the channel. If $V_d > (V_g - V_T)$, we must replace V_d by $V_g - V_T$.

We now introduce an e.m.f. $[S_V(f)\Delta f]^{1/2}$ in series with the gate and equate $S_V(f) = 4kTR_{nf}$, where R_{nf} is the flicker-noise resistance of the device

$$S_V(f) = 4kTR_{nf} = \frac{S_{I_0}(f)}{g_m^2} = \frac{e\mu\bar{\alpha}}{L^2 f} \frac{I_0 V_d}{g_m^2} \quad (7.14)$$

As can be found in any textbook on solid-state devices, we have for an unsaturated MOSFET with a channel of width w and an oxide capacitance

C_{ox} per unit area

$$g_m = \frac{\mu C_{ox} w}{L} V_d; \qquad I_0 = \frac{\mu C_{ox} w}{L}\left[(V_g - V_T)V_d - \frac{1}{2}V_d^2\right] \qquad (7.15)$$

Hence

$$S_V(f) = 4kTR_n = \frac{e\bar{\alpha}}{C_g f}\left(V_g - V_T - \frac{1}{2}V_d\right) \qquad (7.16)$$

where $C_g = C_{ox} wL$ is the device capacitance for zero-drain bias.

This equation holds for a nonsaturated MOSFET. At saturation we must substitute $V_d = V_g - V_T$, so that

$$S_V(f) = 4kTR_n = \frac{e\bar{\alpha}}{2C_g f}(V_g - V_T) \qquad (7.16a)$$

This expression has been experimentally verified by Klaassen,* who found that $S_V(f)$ in saturation varied as $V_g - V_T$, was proportional to the surface-state density and inversely proportional to the device capacitance C_g for zero bias. This does not necessarily mean that (7.16a) is *always* valid, for there may be other flicker-noise mechanisms that cannot be described by this model.

7.2b Noise in Integrated Resistors

Integrated resistors are resistors diffused into a semiconductor substrate. Let the substrate be p type and let an n-type region be diffused into the substrate. If the n region is provided with two contacts we have an integrated resistor isolated from the p-type substrate.

According to (7.5a) the noise in an integrated resistor is

$$S_I(f) = \frac{\alpha I_0^2}{f N_0} \qquad (7.17)$$

where N_0 is the total number of carriers in the sample.

7.3 FLICKER NOISE IN TRANSISTORS

In transistors most of the base current is due to carriers injected into the base and recombining at the base surface or at the surface of the space-

---

*F. M. Klaassen, *I.E.E.E. Trans.*, **ED-18**, 887 (1971).

Flicker Noise in Transistors

charge region. The fluctuating occupancy of oxide traps modulates the (surface) recombination velocity at the base surface and at the surface of the space-charge region, and hence it is not surprising that most of the flicker noise is generated in the base–emitter junction. It must thus be incorporated into the equivalent circuit in the manner shown in Fig. 7.1.

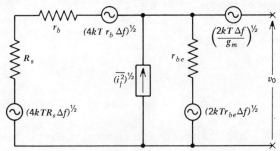

Figure 7.1. Equivalent noise circuit of a common-emitter transistor incorporating flicker noise, here represented by a current generator $(\overline{i_f^2})^{1/2}$ in parallel with the input.

Calculations* show and experiments indicate that the flicker-noise current generator has a mean square value

$$\overline{i_f^2} = B \frac{I_B^\beta}{f} \Delta f \qquad (7.18)$$

where B is a constant and β is of the order of unity.

Since the noise of R_s gives a contribution

$$4kTR_s \Delta f \left(\frac{r_{be}}{R_s + r_b + r_{be}} \right)^2 \qquad (7.19)$$

to $\overline{v_0^2}$ and $\overline{i_f^2}$ gives a contribution

$$\overline{i_f^2} (R_s + r_b)^2 \left(\frac{r_{be}}{R_s + r_b + r_{be}} \right)^2 \qquad (7.20)$$

to $\overline{v_0^2}$, the flicker noise gives a contribution

$$\frac{\overline{i_f^2}}{4kT\Delta f} \frac{(R_s + r_b)^2}{R_s} \qquad (7.21)$$

* A. van der Ziel, *Physica*, **48**, 42 (1970).

to the noise figure F. Hence

$$F = 1 + \frac{r_b}{R_s} + \frac{1}{2}\frac{(R_s+r_b)^2}{R_s r_{be}} + \frac{1}{2g_m R_s}\left(\frac{R_s+r_b+r_{be}}{r_{be}}\right)^2 + \frac{BI_B^\beta}{4kTf}\frac{(R_s+r_b)^2}{R_s} \quad (7.22)$$

At low frequencies the last term in (7.22) predominates and then F has a minimum for $R_s = r_b$. In that case the noise figure becomes, since $r_{be} \gg (R_s + r_b)$,

$$F_{\min} = 2 + \frac{1}{2g_m r_b} + \frac{BI_B^\beta r_b}{fkT} \quad (7.22a)$$

By measuring the noise figure F as a function of source impedance, one can determine r_b.

Plumb and Chenette have written a classical paper on the subject*.

7.4 FLICKER NOISE IN CARBON RESISTORS

Carbon resistors consist of a large number of conducting grains, each making poor contact with the other. The resistance is located in these contacts and this contact resistance fluctuates with time. By passing a current I through the resistor one can detect these fluctuations δR, because they produce a fluctuating e.m.f. $\delta V = I\delta R$, so that

$$S_V(f) = I^2 S_R(f) \quad (7.23)$$

Since $S_R(f)$ has a $1/f$ type noise spectrum, we may write

$$S_V(f) = \frac{AI^2}{f} \quad (7.23a)$$

where A is a constant.

Hence for an amplifier of gain G, bandwidth B, lower cut-off frequency f_2, and upper cut-off frequency $f_1 = (f_2 + B)$

$$\overline{V_{\text{thermal}}^2} = 4kTRG^2 B \quad (7.24)$$

$$\overline{V_{\text{flicker}}^2} = AI^2 G^2 \ln\left(1 + \frac{B}{f_2}\right) \quad (7.24a)$$

*J. L. Plumb and E. R. Chenette, *I.E.E.E. Trans.*, **ED-10**, 304 (1964). For a theoretical discussion of the current dependence of the flicker noise see A. van der Ziel, *Physica*, **48**, 42, (1970).

It thus pays to make f_2 not too small. For example, if $B = 100$ kHz, $f_2 = 100$ Hz, $f_1 = 100,000$ Hz, then $\ln(1 + B/f_2) = 6.9$, whereas if $f_2 = 50$ kHz, $f_1 = 150$ kHz, then $\ln(1 + B/f_2) = 1.08$, so that the total flicker noise is much smaller in the latter case.

To design resistors R with less $1/f$ noise, we take n resistors R in series and n such strings in parallel. Then the total resistance R is the same, but the flicker noise is reduced by a factor n^2.

Proof. If I is the total current, then the current through each string is I/n, so that $S_V(f)$ for each resistor is $1/n^2$ times as large. Since the noises of the n resistors in the string are independent, $S_V(f)$ for each string is $1/n$ times the $S_V(f)$ for the single resistor case. Since the resistance is nR, $S_I(f)$ for a single string is $1/n^3$ times $S_I(f)$ for the single resistor case. However, there are n such strings; hence $S_I(f)$ for the total device is $1/n^2$ times the value for the single resistor case, and the same holds for $S_V(f)$.

To avoid flicker noise, one should avoid carbon resistors in any current-carrying circuit. Instead, one should use wire-wound resistors in all sensitive places in the circuit. However, these resistors have a sizable inductance; if that cannot be tolerated, one should use good metalized resistors. The latter should be checked for flicker noise as shown in Fig. 7.2. If R is the resistor under test, then the flicker noise is tolerable if the meter reading M does not vary when the switch is opened or closed. Any change of meter reading M is due to flicker noise in R.

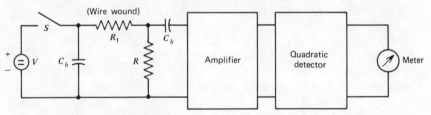

Figure 7.2. Circuit arrangement for testing flicker noise in a resistor R. If the switch S is closed, the output meter shows a larger reading when the resistor has flicker noise.

7.5 GENERATION–RECOMBINATION NOISE IN JFETS

7.5a Generation–Recombination Noise Due to Traps or Impurity Centers in the Channel

If there are electron traps or donor atoms in the channel, and the temperature is so chosen that not all these centers are either occupied or empty, then the number of carriers in the sample will fluctuate due to processes of the type

$$\text{electron + empty trap} \rightleftarrows \text{filled trap or}$$

$$\text{electron + ionized donor} \rightleftarrows \text{neutral donor.}$$

This is the basis of generation–recombination noise in JFETs. A similar consideration holds for p-type channels.

Let a section Δx of the channel at x have ΔN carriers and let $\delta \Delta N$ be the fluctuation in ΔN. Then the corresponding fluctuation δI in the current I_0 is

$$\delta I = -I_0 \frac{\delta \Delta N}{\Delta N}; \quad \text{or} \quad S_I(f) = \frac{I_0^2}{\Delta N^2} S_{\Delta N}(f) \tag{7.25}$$

Now

$$S_{\Delta N}(f) = 4 \overline{\delta \Delta N^2} \frac{\tau}{1+\omega^2\tau^2} = 4\beta \Delta N \frac{\tau}{1+\omega^2\tau^2} \tag{7.26}$$

where we have replaced $\overline{(\delta \Delta N)^2}$ by $\beta \Delta N$ as in Section 7.1; τ is the time constant of the generation–recombination process. Now

$$I_0 = g(V_0) \frac{dV_0}{dx} \tag{7.27}$$

where $g(V_0)$ is the channel conductance for unit length and V_0 is the bias between channel and gate. Here $g(V_0) = (e\mu \Delta N/\Delta x)$, so that

$$S_I(f) = e\mu\beta \frac{I_0^2}{e\mu \Delta N/\Delta x} \frac{1}{\Delta x} \frac{\tau}{1+\omega^2\tau^2} \tag{7.28}$$

Consequently, substituting for I_0 and $g(W_0)$

$$F(x,f) = e\mu\beta I_0 \frac{dV_0}{dx} \frac{\tau}{1+\omega^2\tau^2} \tag{7.29}$$

Substituting into (7.10) yields

$$S_{I_0}(f) = \frac{e\mu\overline{\beta}I_0 V_d}{L^2} \frac{\tau}{1+\omega^2\tau^2} = \frac{4kTR_{ngr}}{g_m^2} \tag{7.30}$$

since $\int_0^L dV_0 = V_d$. We have replaced β by its average value $\overline{\beta}$, to take into account that β may be a slow function of x. Here R_{ngr} is the generation–recombination noise resistance and g_m the transconductance. Equation (7.30) holds for an unsaturated JFET; for a saturated JFET V_d must be replaced by the voltage $V_g - V_T$ for which I_0 saturates.

The time constant τ is often a strong function of the temperature. In most simple cases

$$\tau = \tau_0 \exp\left(\frac{eE_a}{kT}\right) \quad (7.30a)$$

where τ_0 is very small and E_a is the activation energy of the trap or the donor. Typically $E_a = 0.05$–0.20 eV, depending on the trap or donor.

7.5b Generation–Recombination Noise Due to Centers in the Space-Charge Region

In a JFET the junctions are back-biased; as a consequence a recombination center in the space charge region will alternately generate an electron and a hole that are rapidly collected. This causes a fluctuating charge on the center, which gives rise to a locally fluctuating width of the space-charge region, which, in turn, corresponds to a locally fluctuating width of the channel. That finally produces a fluctuating current in the external circuit that can be represented by a noise resistance.

Lauritzen[*] has calculated the effect. He found that in a one-dimensional analysis $S_I(f)$ showed a logarithmic divergence at saturation, but in a two-dimensional analysis $S_I(f)$ remains finite so that a noise resistance $R_n = S_I(f)/(4kTg_m^2)$ can be defined at saturation. This noise resistance is of the form

$$R_n = A(V_g)\frac{\tau}{1+\omega^2\tau^2} \quad (7.31)$$

where τ is the time constant of the centers. Experiments indicate that $A(V_g)$ increases when V_g is made more negative[†]. It would be helpful for design considerations to have a better knowledge of the function $A(V_g)$.

The time constant τ is usually of the order of 10^{-3} sec at room temperature but increases strongly with decreasing temperature T, according to the formula

$$\tau = \tau_0 \exp\left(\frac{eE_a'}{kT}\right) \quad (7.31a)$$

where τ_0 is a small constant and E_a' the activation energy of the center; usually E_a' is approximately equal to half the semiconductor gap width E_g, corresponding to 0.55 V for silicon.

[*] P. O. Lauritzen, *Solid State Electron.*, **8**, 41, 1965.
[†] C. F. Hiatt, Ph. D. Thesis, University of Florida, 1974 (to be published).

8

MEASUREMENT OF SMALL CURRENTS, VOLTAGES, AND CHARGES

8.1 CURRENT MEASUREMENTS

We investigate here how JFETs with floating gate, having a differential input resistance R_g, can be used for measuring very small currents. To that end the current I_0 to be measured is applied to the floating gate (Fig. 8.1a); this results in a change $\Delta V_g = I_0 R_g$ in gate voltage and hence to a change $\Delta I_d = g_m \Delta V_g = g_m R_g I_0$ in the drain current, where g_m is the transconductance of the device, so that the circuit has a current gain $g_m R_g$. This change in current is measured with a critically damped galvanometer.

We saw in Section 3.1b that the minimum detectable current in a critically damped galvanometer of time constant τ_0, defined as the d.c. current that gives the same deflection as the r.m.s. galvanometer deflection due to thermal noise, was given by

$$I_{\min} = \left(\frac{\pi k T}{r \tau_0} \right)^{1/2} \tag{8.1}$$

where r is the resistance of the galvanometer circuit.

We shall now show that our JFET circuit combined with a critically damped galvanometer of time constant τ_0 has a minimum detectable current

$$I_{\min} = \left(\frac{\pi k T}{R_g \tau_0} \right)^{1/2} \tag{8.2}$$

Since $R_g \gg r$, the JFET circuit is much more sensitive than the galvanometer.

Current Measurements

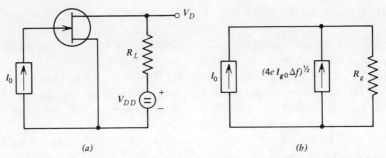

Figure 8.1. (a) A current generator I_0 is connected to a JFET with floating gate; (b) equivalent circuit.

To prove this, we observe that the differential gate resistance R_g shows thermal noise, so that the gate noise current has a mean square value $(4kT\Delta f/R_g)$ (Fig. 8.1b). Since $g_m R_g$ is the current gain of the circuit, the r.m.s. drain current due to gate noise is $g_m R_g (4kT\Delta f/R_g)^{1/2}$.

Now the galvanometer had an input current generator $(4kT\Delta f/r)^{1/2}$ and the resulting r.m.s. noise deflection corresponded to an equivalent input current $(\pi kT/r\tau_0)^{1/2}$ at the galvanometer. Since the noise current $g_m R_g (4kT\Delta f/R_g)^{1/2}$ is $g_m R_g (r/R_g)^{1/2}$ times as large, the corresponding equivalent input current at the galvanometer is

$$g_m R_g \left(\frac{r}{R_g}\right)^{1/2} \left(\frac{\pi kT}{r\tau_0}\right)^{1/2} = g_m R_g \left(\frac{\pi kT}{R_g \tau_0}\right)^{1/2}$$

Now a d.c. current $I_{\min}$ at the *gate* would give rise to a change in drain current $g_m R_g I_{\min}$. Equating this to our noise expression yields (8.2).

If the gate voltage V_g is slightly different from zero,

$$I_g = I_{g0} \exp\left(\frac{eV_g}{kT}\right) - I_{g0}$$

and hence

$$R_g = \left(\frac{\partial I_g}{\partial V_g}\right)^{-1}\bigg|_{V_g=0} = \frac{kT}{eI_{g0}}$$

Substituting into (8.2) yields

$$I_{\min} = \left(\frac{\pi e I_{g0}}{\tau_0}\right)^{1/2} \tag{8.3}$$

How small can we make this? For a good JFET I_{g0} can be as small as 10^{-12} Amp; substituting $e = 1.6 \times 10^{-19}$ Coulomb and taking the time constant of the galvanometer as 5 sec yields $I_{\min} = 3.2 \times 10^{-16}$ Amp, which is several orders of magnitude better than the galvanometer.

For $I_{g0} = 10^{-12}$ Amp and $T = 300°$ K, $R_g = 2.6 \times 10^{10}$ Ω. Assuming that the JFET has an input capacitance $C_g = 5$ pF, the time constant τ_g of the input circuit is 0.13 sec, which is small in comparison with the time constant τ_0 of the galvanometer.

We neglected here the noise resistance R_n of the JFET. In view of the large value of R_g this is well justified.

8.2 DIRECT-CURRENT VOLTAGE MEASUREMENTS

8.2a The Chopper Method

If a small d.c. voltage V_0 is to be measured, it is easiest to transform the voltage into a square wave by means of a mechanical chopper that short-circuits the voltage half the time. The square wave is then stepped up by a step-up transformer, amplified by an a.c. amplifier that is sharply tuned to the chopper frequency f_i and finally detected by a phase sensitive detector driven by a signal of frequency f_i (Fig. 8.2). If T is the period of the chopper, then $f_i = 1/T$. The effective bandwidth B_{eff} of the system can then be made very small by proper choice of the time constant τ of the output circuit of the phase-sensitive detector, and as a consequence the noise limit for V_0 can be made very small.

Let the internal resistance R of the voltage source V_0 under test be relatively small. The thermal noise fed into the transformer is then

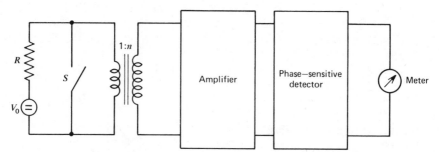

Figure 8.2. A small d.c. voltage V_0 is transformed into a square wave by periodically opening and closing a switch S. The square wave is amplified and detected by a phase-sensitive detector.

Direct-Current Voltage Measurements

$(4kTRB_{\text{eff}})^{1/2}$ for the first half period and zero for the other half; its effective value is thus $(2kTRB_{\text{eff}})^{1/2}$. The a.c. signal produced by the chopper has an amplitude $2V_0/\pi$ at the frequency f_i. The step-up ratio n of the step-up transformer is so chosen that

$$n^2(2kTRB_{\text{eff}}) \gg 4kTR_n B_{\text{eff}}, \quad \text{or} \quad R_n \ll \tfrac{1}{2}n^2 R$$

where R_n is the noise resistance of the a.c. amplifier; the noise of the amplifier can then be neglected. Since the phase sensitive detector measures the a.c. amplitude $2V_0/\pi$, the noise limit is reached when

$$\frac{2V_0}{\pi} = (2kTRB_{\text{eff}})^{1/2} \quad \text{or} \quad V_0 = \frac{\pi}{2}(2kTRB_{\text{eff}})^{1/2} \tag{8.4}$$

As will be shown in the next section

$$B_{\text{eff}} \cong \frac{1}{4\tau} \tag{8.5}$$

where τ is the time constant of the output of the phase-sensitive detector. For $\tau = 5$ sec, $B_{\text{eff}} = 0.050$ Hz.

To find out what the noise limit of the system is, we take $R = 10 \ \Omega$, $B_{\text{eff}} = 0.050$ Hz, and $T = 300°$ K; this yields $V_0 = 10^{-10}$ V, which is quite small.

Good phase-sensitive detectors are commercially available.

8.2b The Phase-sensitive Detector

We shall now show how the phase-sensitive detector operates and we shall demonstrate that if a square noise band of intensity S_0 and center frequency f_i is fed into a phase-sensitive detector driven by a signal of frequency f_i, then the minimum detectable signal is $(S_0 B_{\text{eff}})^{1/2}$ where $B_{\text{eff}} = 1/(4\tau)$ and τ is the time constant of the indicating instrument, taken to be a critically damped galvanometer, in the output of the phase-sensitive detector.

We first demonstrate that the phase-sensistive detector detects the amplitude of the input signal. The phase-sensitive detector is a mixer in which the local oscillator frequency is equal to the input frequency f_i. The time-dependent transconductance of the mixer may then be written

$$g_m(t) = g_{m0} + 2g_{m1}\cos\omega_i t + \text{harmonics} \tag{8.6}$$

where g_{m0} is the average transconductance and g_{m1} the conversion transconductance.

Let now a small a.c. signal $V_i\cos(\omega_i t+\varphi_i)$ of carrier frequency ω_i be applied to the input, then the low-frequency output current is

$$\langle V_i \cos(\omega_i t+\varphi_i) g_m(t)\rangle_{av}$$

where the average is taken over a full cycle of the local oscillator signal. This yields

$$g_{m1}V_i\langle 2\cos(\omega_i t+\varphi_i)\cos\omega_i t\rangle_{av} = g_{m1}V_i\cos\varphi_i \qquad (8.7)$$

which has a maximum absolute value for $\varphi_i=0$ and $\varphi_i=\pm\pi$. The output is thus *phase sensitive*, hence the name of the detector, but the phase adjustment is not critical. We thus see that the detector accurately measures the amplitude V_i, as was previously stated.

To prove the statement about the minimum detectable signal, we consider the response of the phase-sensitive detector to a square noise signal $X(t)$ of bandwidth B, center frequency of f_i, and spectral intensity S_0. As is shown in the next section, we may write

$$X(t) = X_C(t)\cos\omega_i t - X_S(t)\sin\omega_i t \qquad (8.8)$$

where

$$\overline{X^2(t)} = \overline{X_S^2(t)} = \overline{X_C^2(t)} = S_0 B; \qquad \overline{X_C(t)X_C(t+s)} = \overline{X^2}\frac{\sin\pi Bs}{\pi Bs}$$

$$(8.8a)$$

We thus have a carrier signal V_i of frequency f_i with two square noise sidebands of width $B/2$ applied to the detector. In the detection process the carrier beats with the two sidebands; since the noise sidebands are independent, their noise powers add. The detection process consists in multiplying each sideband by g_{m1}^2; hence the spectral intensity of the output noise is

$$S'(f) = 2g_{m1}^2 S_0 \quad\text{for}\quad f<\frac{B}{2}; \qquad S'(f)=0 \text{ otherwise} \qquad (8.9)$$

To normalize the response we put $g_{m1}=1$ mho from here on.

We now pass this signal through an instrument of frequency response $h(f)$; to simplify matters we assume that this response is normalized, that is, $h(0)=1$. Then the deflection A_0 due to the signal is equal to V_i and the

Direct-Current Voltage Measurements

deflection due to the noise has a mean square value

$$\overline{\Delta A^2} = \int_0^{B/2} S'(f) h^2(f) df \cong 2S_0 \int_0^\infty h^2(f) df \qquad (8.10)$$

if $\pi B \tau \gg 1$.

We now bear in mind that the indicating instrument is a critically damped galvanometer of time constant τ. It has a normalized frequency response $h(f) = 1/(1+\omega^2 \tau^2)$, so that

$$\overline{\Delta A^2} \cong 2S_0 \int_0^\infty h^2(f) df = \frac{S_0}{4\tau} = S_0 B_{\text{eff}} \qquad (8.11)$$

Hence $B_{\text{eff}} = 1/(4\tau)$ as stated before, and the minimum detectable signal is

$$(V_i)_{\min} = \left(\overline{\Delta A^2}\right)^{1/2} = (S_0 B_{\text{eff}})^{1/2} \qquad (8.11a)$$

as mentioned earlier.

8.2c Proof of (8.8), (8.8a), and (8.9)

Let for $0 \leqslant t \leqslant T$

$$X(t) = \sum_{n=1}^\infty c_n \cos(\omega_n t - \varphi_n), \quad \text{with} \quad \omega_n = \frac{2\pi n}{T}$$

Put $\cos(\omega_n t - \varphi_n) = \cos[(\omega_n - \omega_i)t - \varphi_n + \omega_i t]$ and expand. Then

$$X(t) = X_C(t) \cos \omega_i t - X_S(t) \sin \omega_i t,$$

with

$$X_C(t) = \sum_{n=1}^\infty c_n \cos[(\omega_n - \omega_i)t - \varphi_n];$$

$$X_S(t) = \sum_{n=1}^\infty c_n \sin[(\omega_n - \omega_i)t - \varphi_n]$$

This proves (8.8).

Since the elements of the ensemble under consideration have random phase, $\overline{\cos^2[(\omega_n - \omega_i)t - \varphi_n]} = \frac{1}{2}$, and so on, so that

$$\overline{X_C^2(t)} = \overline{X_S^2(t)} = \sum_{n=1}^\infty \overline{\tfrac{1}{2} c_n^2} = \overline{X^2(t)} = S_0 B$$

To prove the last part of (8.8a) we put

$$\overline{X_C(t)X_C(t+s)} = \sum_{j=1}^{\infty}\sum_{k=1}^{\infty}\overline{c_j\cos[(\omega_j-\omega_i)t-\varphi_j]c_k\cos[(\omega_k-\omega_i)(t+s)-\varphi_k]}$$

$$= \sum_{j=1}^{\infty}\overline{c_j^2\,\cos[(\omega_j-\omega_i)t-\varphi_j]\cos[(\omega_j-\omega_i)(t+s)-\varphi_j]}$$

$$= \frac{1}{2}\sum_{j=1}^{\infty}\overline{c_j^2}\,\cos(\omega_j-\omega_i)s = \int_0^{\infty} S_x(f)\cos(\omega-\omega_i)s\,df$$

$$= \frac{S_0}{2\pi s}\int_{-B/2}^{B/2}\cos(\omega-\omega_i)s\,d[(\omega-\omega_i)s] = S_0 B\,\frac{\sin\pi Bs}{\pi Bs}$$

The same expression holds for $\overline{X_S(t)X_S(t+s)}$.

We finally give a rigorous proof of (8.9). The low-frequency output-noise current is

$$I(t) = \langle g_m(t)X(t)\rangle_{av} = 2g_{m1}\langle X_C(t)\cos^2\omega_i t\rangle_{av} = g_{m1}X_C(t)$$

since $\langle\cos^2\omega_i t\rangle_{av} = \tfrac{1}{2}$ when the averaging is performed over a full local oscillator cycle.

We now apply the Wiener–Khintchine theorem to this random signal. The spectral intensity $S'(f)$ is then

$$S'(f) = 4g_{m1}^2\int_0^{\infty}\overline{X_C(t)X_C(t+s)}\cos 2\pi fs\,ds$$

$$= 4g_{m1}^2 S_0 B\int_0^{\infty}\left(\frac{\sin\pi Bs}{\pi Bs}\right)\cos 2\pi fs\,ds$$

$$= 2g_{m1}^2 S_0 \text{ for } f<(B/2) \text{ and zero otherwise.}$$

8.3 MEASUREMENT OF SMALL CHARGES

Suppose by some means we collect a small charge Q on a capacitor C, then the capacitor voltage $V_C(t)$ rises. The capacitor is now connected to an amplifier having a low-noise JFET as its first stage, and filtered by a high-pass filter with low-frequency cut-off f_1 and a low-pass filter with high-frequency cut-off f_2, where $f_2 > f_1$; this transforms $V_C(t)$ into a pulse. By proper choice of f_1 and f_2 the signal-to-noise ratio can be optimized and the noise can be expressed as the minimum number N_{min} of electron charges that can be detected. We shall see that N_{min} is of the order of 50 electron charges at room temperature.

To simplify the calculation we shall assume that $V_C(t)$ is a ramp

$$\frac{Q}{C_{tot}}\left(\frac{t-t_0}{\tau}\right) \tag{8.12}$$

for $0 \leqslant t-t_0 \leqslant \tau$, followed by a very slow decay. Here C_{tot} is the total capacitance of the input circuit. The high-pass filter is now replaced by a differentiating circuit that transforms the ramp into a pulse and the low-pass filter used has a frequency response $(1+jf/f_0)^{-2}$, obtained by connecting two RC filters in cascade; this filters out the high-frequency noise. The only unknown parameter is now f_0; by evaluating the signal-to-noise ratio at the output of the system and differentiating it with respect to f_0, we can obtain the optimum signal-to-noise ratio and express it in terms of electron charges (Fig. 8.3).

8.3a Signal Considerations

If the first amplifier stage is a JFET with floating gate, gate-source capacitance C_{gs}, gate-drain capacitance C_{gd}, transconductance g_m, and R_L is the load resistance in the drain, then

$$C_{tot} = C + C_{gs} + C_{gd}(1 + g_m R_L) \tag{8.12a}$$

where $C_{gd}g_m R_L$ represents the Miller effect capacitance.

The signal (8.12) is now amplified by a wide band amplifier of gain G and differentiated to produce a pulse of height

$$\frac{GQ}{C_{tot}\tau} \tag{8.12b}$$

of duration τ, starting at $t = t_0$.

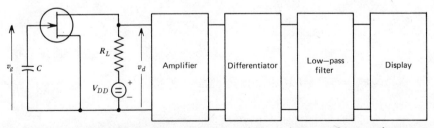

Figure 8.3. A charge Q, collected on a capacitor C during a time τ, produces a voltage ramp across C that is amplified, differentiated, passed through a low-pass filter, and detected.

This pulse is fed into the filter of frequency response $(1+jf/f_0)^{-2}$, resulting in a transient

$$\frac{GQ}{C_{tot}\tau}\{1+\omega_0(t-t_0-\tau)-[1+\omega_0(t-t_0)]\exp(-\omega_0\tau)\}$$

$$\cdot\exp[-\omega_0(t-t_0-\tau)] \quad (8.13)$$

where $\omega_0=2\pi f_0$. This pulse has a peak at

$$\omega_0(t-t_0-\tau)=\omega_0\tau\frac{\exp(-\omega_0\tau)}{1-\exp(-\omega_0\tau)} \quad (8.13a)$$

If $\omega_0\tau\ll 1$ the right-hand side of (8.13a) is unity; we assume for sake of simplicity that this inequality is satisfied.

The peak in the output response is then obtained for $\omega_0(t-t_0-\tau)=1$. Substituting into (8.13) and bearing in mind $\omega_0\tau\ll 1$ yields a pulse height

$$\frac{GQ}{C_{tot}}\frac{2\pi f_0}{\varepsilon} \quad (8.14)$$

where ε is the basis of natural logarithms.

8.3b Signal-to-Noise Ratio

To calculate the noise, we observe that we have two noise currents, i_g in the gate circuit and i_n in the drain circuit. For a floating gate

$$\overline{i_g^2}=4eI_{g0}\Delta f: \qquad \overline{i_n^2}=4kTR_n\Delta f g_m^2 \quad (8.15)$$

We now have for the drain current

$$i_d=g_m v_g+i_n \quad (8.16)$$

$$i_g+(v_d-v_g)j\omega C_{dg}=v_g j\omega(C+C_{gs}) \quad (8.17)$$

$$v_d=-i_d R_L=-g_m R_L v_g-i_n R_L \quad (8.18)$$

Solving for v_d yields after some manipulations

$$v_d=\frac{[g_m i_g+i_n j\omega(C+C_{gs}+C_{gd})]R_L}{j\omega(C+C_{gs}+C_{gd}g_m R_L)} \quad (8.19)$$

Measurement of Small Charges

Introducing an equivalent current generator i_{eq} at the input by equating

$$v_d = \frac{g_m i_{eq} R_L}{j\omega(C + C_{gs} + C_{gd} + C_{gd}g_m R_L)} \quad (8.20)$$

we obtain

$$i_{eq} = i_g + \frac{i_n}{g_m} j\omega(C + C_{gs} + C_{gd}) \quad (8.20a)$$

We thus see that the Miller effect does not affect i_{eq}. Introducing an equivalent saturated diode current I_{eq} by equating $\overline{i_{eq}^2} = (2eI_{eq}\Delta f)$ and substituting for $\overline{i_g^2}$ and $\overline{i_n^2}$ from (8.15) yields

$$I_{eq} = 2I_{g0} + \frac{2kT}{e} R_n \omega^2 (C + C_{gs} + C_{gd})^2 \quad (8.21)$$

Now I_g is a constant and $R_n = (R_{n0} + A/f)$, where R_{n0} is the thermal noise of the device and A/f is a possible flicker noise. We may thus write

$$I_{eq} = a + bf + cf^2 \quad (8.22)$$

where

$$a = 2I_{g0}; \quad b = \frac{2kT}{e}(2\pi)^2 A(C + C_{gs} + C_{gd})^2;$$

$$c = \frac{2kT}{e}(2\pi)^2 R_{n0}(C + C_{gs} + C_{gd})^2 \quad (8.22a)$$

The noise in a frequency interval df produces an equivalent input voltage of mean square value

$$2e \, df \frac{(a + bf + cf^2)}{\omega^2 C_{tot}^2} \quad (8.23)$$

This is amplified (i.e., multiplied by G^2), differentiated (i.e., multiplied by ω^2), and filtered, [i.e., multiplied by $(1 + f^2/f_0^2)^{-2}$]. The total mean square noise voltage is thus

$$\frac{2eG^2}{C_{tot}^2} \int_0^\infty \frac{(a + bf + cf^2) df}{(1 + f^2/f_0^2)^2} = \frac{2eG^2}{C_{tot}^2} \left[\frac{\pi}{4} af_0 + \frac{1}{2} bf_0^2 + \frac{\pi}{4} cf_0^3 \right] \quad (8.23a)$$

Dividing by the square of (8.14) gives the noise-to-signal power ratio as

$$e\frac{2\varepsilon^2}{4\pi^2 Q^2}\left[\frac{\pi}{4}\frac{a}{f_0}+\frac{1}{2}b+\frac{\pi}{4}cf_0\right] \quad (8.24)$$

This is a minimum for $f_0=(a/c)^{1/2}$ and the minimum is

$$\frac{e^2\varepsilon^2}{4\pi^2 Q^2}\left[\frac{b+\pi(ac)^{1/2}}{e}\right] \quad (8.24a)$$

The minimum value of Q is found by equating this expression to unity

$$Q_{\min}=eN_{\min}=\frac{\varepsilon e}{2\pi}\left[\frac{b+\pi(ac)^{1/2}}{e}\right]^{1/2} \quad (8.24b)$$

To find out the magnitude of $N_{\min}$ we assume that we have a small JFET with $I_{g0}=0.50\times 10^{-12}$ A, $(C+C_{gs}+C_{gd})=5$ pF, $R_{n0}=10^4$ Ω, and $A=0$ (the JFET has practically no flicker noise). Then $a=10^{-12}$, $b=0$, and $c=5\times 10^{-19}$; hence

$$f_0=1400 \text{ Hz}, \quad N_{\min}=51$$

Since the term $\pi(ac)^{1/2}$ is the significant term, improvement can only be obtained by lowering I_{g0} and R_{n0}. This can be achieved by cooling the device. In this manner values of $N_{\min}$ less than 10 have been obtained. This is very important for nuclear particle detectors.

It is beyond the scope of this discussion to go into details of the practical implementation of this outline. We refer to commercial manuals about this problem.

REFERENCES

G. Dearnaley and D. C. Northrop, *Semiconductor Counters for Nuclear Reactions*, Wiley, New York, 1966.

G. Bertolini and A. Coche (eds.), *Semiconductor Detectors*, Wiley-Interscience, New York; North-Holland, Amsterdam, 1968.

9

THERMAL RADIATION DETECTORS

In this chapter we discuss radiation detectors in which the radiation produces heat, the heat changes a parameter of the device under study, and this parameter is detected. In particular we are discussing the thermocouple (Section 9.2) and the resistive bolometer (Section 9.3). The pyroelectric detector and the capacitive bolometer, which also belong to this class, are discussed in Chapter 12.

To discuss radiation detectors, the concepts of technical sensitivity, noise-equivalent power (NEP), and detectivity are introduced (Section 9.1a). These concepts are much more general and apply to radiation detectors that are not of the thermal detector class.

The limiting noise source in thermal radiation detectors is the spontaneous temperature fluctuation in the detector (Section 9.1b). This limit is not always reached, however.

Another noise source is the noise in the radiation itself. One has to distinguish here between the cases of thermal radiation (Section 9.1c) and of laser radiation (Section 9.1d).

9.1 GENERAL CONSIDERATIONS

9.1a Technical Sensitivity, Noise Equivalent Power, and Detectivity

We first define technical sensitivity. Let a radiation detector receive modulated (e.g., chopped) radiation. Let a radiant power P produce a reading R of the instrument. We then call

$$C = \frac{R}{P} \qquad (9.1)$$

the *technical sensitivity* of the detector. For example, a detector produces a voltage V then

$$C = C_V = \frac{V}{P} \tag{9.1a}$$

is expressed in V/W. Or a detector produces a current I, then

$$C = C_I = \frac{I}{P} \tag{9.1b}$$

it is expressed in A/W.

We now define the NEP. Let the noise of the detector plus indicating instrument have a spectral intensity $S_R(f)$. We then define the NEP, P_{eq}, by the definition

$$CP_{eq} = [S_R(f)]^{1/2} \quad \text{or} \quad P_{eq} = \frac{[S_R(f)]^{1/2}}{C} \tag{9.2}$$

It is expressed in W/Hz$^{1/2}$, and is the power that gives the same reading as the noise per unit bandwidth. It can always be defined; a good detector is one that has the smallest P_{eq}. For example, if the detector produces a voltage and has a spectral intensity $S_V(f)$ in this voltage, then

$$P_{eq} = \frac{[S_V(f)]^{1/2}}{C_V} \tag{9.2a}$$

Or, if the detector produces a current, and has a spectral intensity $S_I(f)$ in this current, then

$$P_{eq} = \frac{[S_I(f)]^{1/2}}{C_I} \tag{9.2b}$$

Usually P_{eq} is proportional to the square root of the detector area A. If the detector area A can be freely chosen, one can then normalize the value of P_{eq} to a unit area, by setting

$$P_{eq}^* = \frac{P_{eq}}{A^{1/2}} \tag{9.3}$$

which is expressed in W/(Hz$^{1/2}$cm).

The best detector type is the one with the smallest P_{eq}^*. Since one wants to identify a good detector by a large number, one introduces the *detectivity* D^* by

$$D^* = \frac{1}{P_{eq}^*} = \frac{A^{1/2}}{P_{eq}} \tag{9.3a}$$

General Considerations

It is expressed in cmHz$^{1/2}$/W and is a measure for the intrinsic properties of the detector system.

In some detectors the area A cannot be freely chosen but is prescribed by design considerations. In such a case D^* and P^*_{eq} have no meaning and only P_{eq} is a meaningful parameter.

9.1b Temperature-Fluctuation Noise

In many cases the noise limit of a detector is set by the spontaneous temperature fluctuations of the detector. This is the case for any detector that transforms a modulated radiant power P into a temperature variation ΔT. In that case we put

$$\Delta T = C_T P \tag{9.4}$$

and call C_T the technical sensitivity of the detector.

We now must compare this with the spontaneous temperature fluctuations of the device. Assuming that the device has an effective area A, a heat capacity C_H, and heat-loss conductance g_H and g'_H at the front and the back, respectively, then

$$S_{\Delta T}(f) = \frac{4kT^2(g_H + g'_H)}{(g_H + g'_H)^2 + \omega^2 C_H^2} \tag{9.5}$$

[cf. (5.44a)].

We now evaluate C_T. The heat equation of the system is

$$C_H \frac{d\Delta T}{dt} + (g_H + g'_H)\Delta T = \eta P_1 \exp(j\omega t) \tag{9.6}$$

where η is the emissivity of the front face. Substituting $\Delta T = \Delta T_0 \exp(j\omega t)$

$$(j\omega C_H + g_H + g'_H)\Delta T_0 = \eta P_1 \tag{9.6a}$$

Hence

$$(\Delta T_0)_{\text{r.m.s.}} = \frac{\eta (P_1)_{\text{r.m.s.}}}{\left[(g_H + g'_H)^2 + \omega^2 C_H^2\right]^{1/2}} \tag{9.7}$$

from which C_T can be determined.

We now define P_{eq} by equating

$$(\Delta T_0)_{\text{r.m.s.}} = \left[S_{\Delta T}(f)\right]^{1/2}$$

and obtain

$$P_{eq} = \frac{\left[4kT^2(g_H + g'_H)\right]^{1/2}}{\eta} \tag{9.8}$$

Now if all heat loss is by radiation, and η' is the emissivity of the back face

$$g_H = \eta \cdot 4\sigma T^3 A, \qquad g'_H = \eta' \cdot 4\sigma T^3 A \tag{9.8a}$$

where $\sigma = 5.67 \times 10^{-12}$ W cm^{-2} deg^{-4} is the Stephan–Boltzmann constant. The proof is simple. If the surrounding medium has a temperature T and the detector has a temperature $T + \Delta T$, then the heat lost by radiation is given by the Stephan–Boltzmann law as

$$\eta\sigma(T+\Delta T)^4 A - \eta\sigma T^4 A = 4\eta\sigma T^3 A \, \Delta T = g_H \Delta T$$

from which (9.8a) follows.
Therefore

$$P_{eq} = \frac{\left[16 A\sigma k T^5 (\eta + \eta')\right]^{1/2}}{\eta} \tag{9.9}$$

In the ideal case $\eta = 1$ and $\eta' = 0$, or

$$P_{eq} = (16 A\sigma k T^5)^{1/2} \tag{9.9a}$$

whereas P_{eq} is a factor $2^{1/2}$ larger in the case $\eta = \eta' = 1$. Consequently

$$D^* = \frac{A^{1/2}}{P_{eq}} = \frac{\eta}{\left[16\sigma k T^5 (\eta + \eta')\right]^{1/2}} \tag{9.9b}$$

Example. $A = 1$ mm^2, $T = 300°$ K. Substituting the values for σ and k yields $P_{eq} = 5.5 \times 10^{-12}$ W/Hz$^{1/2}$ for $\eta = 1$, $\eta' = 0$, and $P_{eq} = 7.8 \times 10^{-12}$ W/Hz$^{1/2}$ for $\eta = \eta' = 1$, so that the values of D^* are 1.8×10^{10} and 1.3×10^{10} cm Hz$^{1/2}$/W, respectively.

If the detector is at a temperature T_1 and the environment is at a temperature T_0, then, in analogy with (9.9)

$$P_{eq} = \frac{\left[8 A\sigma k T_1^5 (\eta + \eta') + 8 A\sigma k T_0^5 (\eta + \eta')\right]^{1/2}}{\eta} \tag{9.10}$$

General Considerations

This means that somewhat better results can be obtained by cooling the detector. Still better results can be obtained by protecting the detector with a heat shield that cuts out most of the room temperature background radiation. In that case (9.10) may be written

$$P_{eq} = \frac{\left[8A\sigma k T_1^5 (\eta + \eta') + S \cdot 8A\sigma k T_0^5 (\eta + \eta')\right]^{1/2}}{\eta} \quad (9.10a)$$

where S is a shielding factor.

In the limiting case the heat loss is determined by heat conduction through the contact wires (Section 9.2).

9.1c Fluctuations in Black-body Radiation

We shall now show that if n quanta of black-body radiation are received per second, then

$$\operatorname{var} n = \frac{\bar{n}}{1 - \exp(-h\nu/kT)} \quad (9.11)$$

where T is the absolute temperature of the black-body and $h\nu$ is the energy of the emitted quanta. In other words, $\operatorname{var} n$ can be larger than $\bar{n}$ (super-Poissonian noise).

We first investigate a single black-body mode. An optical cavity has a set of standing wave patterns of frequency ν and each standing wave pattern can have two independent, perpendicular directions of polarization. A standing wave pattern with a particular direction of polarization is called a *black-body mode*. Such a black-body mode can be treated as a harmonic oscillator of frequency ν, and according to quantum theory its energy E is quantized

$$E_v = h\nu(v + \tfrac{1}{2})(v = 0, 1, 2 \ldots)$$

where h is Planck's constant, and $\tfrac{1}{2}h\nu$ is called the zero-point energy. This zero-point energy is an additive constant that we shall neglect. If the quantum number v is larger than zero, it is said that there are v quanta in the mode.

According to Boltzmann's theorem, the probability that there are v quanta in the mode is proportional to $\exp(-vh\nu/kT)$ and hence it may be written as

$$C\exp\left(-v\frac{h\nu}{kT}\right)$$

where the normalization factor C must be so chosen that

$$C\sum_v \exp\left(-\frac{vh\nu}{kT}\right) = 1 \quad \text{or} \quad C = 1 - \exp\left(-\frac{h\nu}{kT}\right)$$

We now determine

$$\bar{v} = C\sum_v v\exp\left(-v\frac{h\nu}{kT}\right); \quad \overline{v^2} = C\sum_v v^2 \exp\left(-v\frac{h\nu}{kT}\right)$$

To that end we put $(h\nu/kT) = x$ and write

$$\sum_v \exp(-vx) = \frac{1}{1-\exp(-x)}$$

We now differentiate both sides with respect to x. This yields

$$\sum_v -v\exp(-vx) = -\frac{\exp(-x)}{[1-\exp(-x)]^2},$$

or

$$\sum_v v\exp(-vx) = \frac{\exp(-x)}{[1-\exp(-x)]^2}$$

Differentiating both sides once more with respect to x yields

$$\sum_v v^2 \exp(-vx) = \frac{\exp(-x)}{[1-\exp(-x)]^2} + 2\frac{[\exp(-x)]^2}{[1-\exp(-x)]^3}$$

$$= \frac{\exp(-x) + [\exp(-x)]^2}{[1-\exp(-x)]^3}$$

General Considerations

Therefore, multiplying by C yields

$$\bar{v} = \frac{\exp(-x)}{1-\exp(-x)}; \qquad \overline{v^2} = \frac{\exp(-x) + [\exp(-x)]^2}{[1-\exp(-x)]^2}$$

$$\operatorname{var} v = \overline{v^2} - (\bar{v})^2 = \frac{\exp(-x)}{[1-\exp(-x)]^2} = \frac{\bar{v}}{1-\exp(-h\nu/kT)} \qquad (9.11a)$$

as had to be proved.

Hence (9.11) is thus true for a single mode. If we have a large number of modes, with $n = \sum p_i$ quanta hn those modes, then $\bar{n} = \sum_i \bar{v}_i$ and

$$\operatorname{var} n = \frac{\sum_i \bar{v}_i}{1-\exp(-h\nu/kT)} = \frac{\bar{n}}{1-\exp(-h\nu/kT)}$$

Therefore, (9.11) is true for a large number of modes around ν.

Since a mode with energy $v h \nu$ emits v photons practically simultaneously, the number of *emitted* quanta satisfies the same formula. Unless part of the radiation is intercepted, the number of received quanta also satisfies (9.11).*

9.1d Fluctuations in Laser Radiation

A laser far above threshold can be considered as a generator of a signal with constant amplitude. It can be shown that the emitted radiation obeys Poisson statistics, namely $\operatorname{var} n = \bar{n}$.

To explain this, we consider an ideal laser with zero optical losses. Then the rate n of *emission* of photons is equal to the *pumping* rate W (either optical pumping or pumping by any other means), whereby atoms are transferred from a lower to a higher energy level. Now the pumping consists of a series of independent random events, so that $\operatorname{var} W = \overline{W}$; since $n = W$, we have $\operatorname{var} n = \bar{n}$.

Next we allow losses to occur. If n is the rate of *produced* quanta, then $\operatorname{var} N = \overline{N}$, as before. If now the part λ of the produced quanta is actually emitted, we have from the variance theorem for the emission rate n

$$\bar{n} = \overline{N}\lambda; \qquad \operatorname{var} n = \lambda^2 \operatorname{var} N + \overline{N}\lambda(1-\lambda) = \overline{N}\lambda = \bar{n}$$

so that the *emitted* quanta also show full shot noise.

---

*For an experimental verification of (9.11) see: G. W. Kattke and A. van der Ziel, *Physica*, **49**, 461, 1970.

We can also use the following reasoning. An arbitrary spectral line gives *at least* shot noise of the quanta contained in the line. In addition, *beats* occur between the frequencies within the line. This is called *wave-interaction noise*; it gives an additional contribution to var n. However, a single-mode laser far above threshold is a single-frequency oscillator that produces no beats and hence no wave-interaction noise. Hence var $n = \bar{n}$.

Close to threshold there is some extra noise due to the spontaneous emission f quanta. The oscillation produces a carrier signal; the spontaneous emission noise produces two sidebands, one on each side of the carrier frequency ν of the laser. Upon detection the carrier beats with the two noise sidebands and produces an additional low-noise component. Near threshold the effect is quite pronounced but far above threshold it drowns in the shot noise.*

9.2 THE THERMOCOUPLE DETECTOR

Radiation is received by a thin blackened plate of area A to which two thermocouple wires 1 and 2 of different material are connected. The temperature variation ΔT produced by the radiation gives rise to a thermo e.m.f. $s\Delta T$ in the thermocouple. This e.m.f. is measured with a meter of internal resistance R_i (Fig. 9.1).

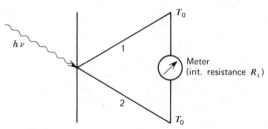

Figure 9.1. Thermocouple with a receiving plate S receives quanta that heat the receiving plate and produce a thermo e.m.f. between the wires 1 and 2 that are connected to a meter M with internal resistance R_i. The end points of the wires are kept at the temperature T_0.

We start by writing down the basic equations of the system. Let the wires 1 and 2 have specific resistances ρ_1 and ρ_2, heat conductances k_1 and k_2, and let $a_1 = (A_1/L_1)$ and $a_2 = (A_2/L_2)$ be the ratio of cross-sectional area A over the length of L of each wire. The total resistance r of the

*For references see the following review paper: A. van der Ziel, *Proc. IEEE*, **58**, 1178 (1970).

The Thermocouple Detector

circuit is then

$$r = \frac{\rho_1}{a_1} + \frac{\rho_2}{a_2} + R_i \tag{9.12}$$

and the total heat-loss conductance is

$$g_H = 4\eta A \sigma T^3 + k_1 a_1 + k_2 a_2 \tag{9.13}$$

where the first term describes the heat loss due to radiation. We shall assume $\eta = 1$ in our discussion.

The general set of equations describing the system is

$$C_H \frac{d\Delta T}{dt} + g_H \Delta T + sTi = P_1 \exp(j\omega t) + H(t) \tag{9.14}$$

$$ir - s\Delta T = e_0 \exp(j\omega t) + E(t) \tag{9.15}$$

Here $H(t)$ is the random source function describing the temperature-fluctuation noise and $E(t)$ the random-source function describing thermal noise in the electrical circuit, whereas $P_1 \exp(j\omega t)$ and $e_0 \exp(j\omega t)$ are, respectively, the radiation signal and electric signal needed for determining the technical sensitivity and the device impedance. Finally i is the current and sTi the heat loss rate due to Peltier effect, where s is the Seebeck coefficient.

9.2a Technical Sensitivity and Device Impedance

We first evaluate the technical sensitivity. To do so, we ignore $H(t)$ and $E(t)$ and put $e_0 = 0$. We observe that in that case $i = (s\Delta T/r)$, and hence the Peltier heat loss is $sTi = s^2 T \Delta T / r$. Substituting $\Delta T = \Delta T_0 \exp(j\omega t)$ yields for $\eta = 1$

$$\left(j\omega C_H + g_H + \frac{s^2 T}{r} \right) \Delta T_0 = P_1; \quad \Delta T_0 = \frac{P_1}{j\omega C_H + g_H + s^2 T / r} \tag{9.16}$$

or

$$i_0 = \frac{s\Delta T}{r} = \frac{s/r}{j\omega C_H + g_H + s^2 T / r} P_1 \tag{9.16a}$$

so that the technical sensitivity of the detector is

$$\frac{s/r}{\left[\omega^2 C_H^2 + \left(g_H + s^2 T / r \right)^2 \right]^{1/2}} \tag{9.17}$$

For $\omega^2 C_H^2 \ll (g_H + s^2T/r)^2$ this is equal to

$$\frac{s/r}{g_H + s^2T/r} \tag{9.17a}$$

so that the technical sensitivity is here independent of frequency. For $\omega^2 C_H^2 \gg (g_H + s^2T/r)^2$, (9.17) may be written

$$\frac{s/r}{\omega C_H} \tag{9.17b}$$

which is inversely proportional to the frequency.

We next evaluate the impedance Z_{in} seen by $e_0 \exp(j\omega t)$. To that end we put $P_1 = 0$, ignore $H(t)$ and $E(t)$, and put $\Delta T = \Delta T_0 \exp(j\omega t)$ and $i = i_0 \exp(j\omega t)$. This yields

$$(j\omega C_H + g_H)\Delta T_0 + sTi_0 = 0; \quad i_0 r - s\Delta T_0 = e_0 \tag{9.18}$$

or

$$\Delta T_0 = -\frac{sT}{j\omega C_H + g_H} i_0 \tag{9.18a}$$

Hence the input impedance is

$$Z_{in} = \frac{e_0}{i_0} = r - \frac{s\Delta T_0}{i_0} = r + \frac{s^2 T}{j\omega C_H + g_H} \tag{9.19}$$

This corresponds to

$$Z_{in} = r + \frac{s^2 T}{g_H} \tag{9.19a}$$

at low frequencies ($\omega^2 C_H^2 \ll g_H^2$), which is purely resistive. For $\omega^2 C_H^2 \gg g_H^2$ (high frequencies)

$$Z_{in} = r + \frac{s^2 T}{j\omega C_H} \tag{9.19b}$$

which consists of the resistance r and a capacitance term in series.

9.2b Noise and P_{eq}

We now calculate the noise. To do so we put $P_1 = 0$ and $e_0 = 0$ and obtain

$$C_H \frac{d\Delta T}{dt} + g_H \Delta T + sTi = H(t)$$

$$ir - s\Delta T = E(t) \quad \text{or} \quad i = \frac{s\Delta T}{r} + \frac{E(t)}{r}$$

The Thermocouple Detector

Hence we obtain the Langevin equation

$$C_H \frac{d\Delta T}{dt} + \left(g_H + \frac{s^2 T}{r}\right)\Delta T = H(t) - \frac{sT}{r} E(t) \qquad (9.20)$$

Substituting

$$\Delta T = \sum_n \Delta T_n \exp(j\omega_n t), \qquad H(t) = \sum_n h_n \exp(j\omega_n t);$$

$$E(t) = \sum_n e_n \exp(j\omega_n t)$$

yields

$$\left(j\omega_n C_H + g_H + \frac{s^2 T}{r}\right)\Delta T_n = h_n - \frac{sT}{r} e_n;$$

$$\Delta T_n = \frac{h_n - (sT/r)e_n}{j\omega_n C_H + g_H + s^2 T/r} \qquad (9.20a)$$

or

$$i_n = \frac{s}{r}\Delta T_n + \frac{e_n}{r} = \frac{(s/r)h_n + (j\omega_n C_H + g_H)e_n/r}{j\omega_n C_H + g_H + s^2 T/r} \qquad (9.21)$$

Since $S_H(f) = 4kT^2 g_H$ and $S_E(f) = 4kTr$,

$$S_I(f) = \frac{(s^2/r^2)4kT^2 g_H + (4kT/r)(\omega^2 C_H^2 + g_H^2)}{\omega^2 C_H^2 + (g_H + s^2 T/r)^2} \qquad (9.22)$$

Hence the circuit e.m.f. has a spectral intensity $S_V(f) = S_I(f)|Z_{in}|^2$. However, Z_{in} is given by (9.19), and hence

$$S_V(f) = \frac{(s^2/r^2)4kT^2 g_H + (4kT/r)(\omega^2 C_H^2 + g_H^2)}{\omega^2 C_H^2 + (g_H + s^2 T/r)^2} \cdot r^2 \frac{(g_H + s^2 T/r)^2 + \omega^2 C_H^2}{(\omega^2 C_H^2 + g_H^2)}$$

$$= 4kT\left[r + \frac{s^2 T g_H}{\omega^2 C_H^2 + g_H^2}\right] = 4kT\,\mathrm{Re}(Z_{in}) \qquad (9.23)$$

corresponding to thermal noise of the real part of the input impedance. This is as expected, for we are here dealing with a device in equilibrium at a temperature T; hence the circuit should indeed show thermal noise.

We now evaluate P_{eq} by equating

$$(i_0)_{\mathrm{r.m.s.}} = [S_I(f)]^{1/2}$$

This yields

$$\frac{sP_{eq}/r}{\left[\omega^2 C_H^2 + (g_H + s^2 T/r)^2\right]^{1/2}} = \frac{\left[(s^2/r^2)4kT^2 g_H + (4kT/r)(\omega^2 C_H^2 + g_H^2)\right]^{1/2}}{\left[\omega^2 C_H^2 + (g_H + s^2 T/r)\right]^{1/2}}$$

or

$$P_{eq} = \left[4kT^2 g_H + \frac{4kTr}{s^2}(\omega^2 C_H^2 + g_H^2)\right]^{1/2} \quad (9.24)$$

At low frequencies the second term usually predominates already, so that it certainly predominates at high frequencies. We thus have at high frequencies ($\omega^2 C_H^2 \gg g_H^2$)

$$P_{eq} = (4kTr)^{1/2} \frac{\omega C_H}{s} \quad (9.24a)$$

This is optimized by making r as small as possible. Furthermore, P_{eq} is proportional to the modulation frequency ω, so ω should be kept small.

At low frequencies ($\omega^2 C_H^2 \ll g_H^2$), however,

$$P_{eq} = (4kT^2)^{1/2} \left[g_H + \frac{r}{s^2 T} g_H^2\right]^{1/2} \quad (9.24b)$$

The optimization of this expression is more tedious. First R_i is so chosen that the meter is matched to the thermocouple for maximum power transfer. That means

$$R_i = \frac{\rho_1}{a_1} + \frac{\rho_2}{a_2} \quad \text{or} \quad r = 2\left(\frac{\rho_1}{a_1} + \frac{\rho_2}{a_2}\right) \quad (9.25)$$

We observe that g_H is given by (9.13). Next a_1 and a_2 are so chosen that P_{eq} is minimized. We obtain after an elementary but lengthy calculation

$$P_{eq} = (16A\sigma kT^5)^{1/2} \left[y + (1+y^2)^{1/2}\right] \quad (9.26)$$

where

$$y = \left(\frac{2k_0 \rho_0}{s^2 T}\right)^{1/2}; \quad k_0 = k_1 + k_2 \left(\frac{k_1 \rho_2}{k_2 \rho_1}\right)^{1/2}; \quad \rho_0 = \rho_1 + \rho_2 \left(\frac{k_2 \rho_1}{k_1 \rho_2}\right)^{1/2}$$

(9.26a)

The first term between brackets corresponds to thermal fluctuation noise

The Resistive Bolometer

limit,* whereas the factor $y+(1+y^2)^{1/2}$ indicates how far we are away from that limit.

The following table indicates how large this factor is for different thermocouple materials:

Thermocouple	$y+(1+y^2)^{1/2}$
Constantan–manganese	50
Bismuth–antimony	25
Antimony–bismuth–antimony alloy	9

Stevens[†] gives a list of evaporated thermocouples. The results are shown in the following table

Characteristic	1×1 mm	0.25×0.25 mm	2 mm diameter	0.12 mm×0.12 mm
Responsivity (V/W, vacuum)	50	220	160	280
Time constant τ_d (μs, vacuum)	100	75	150	13
Impedance Z (kΩ)	6.3	10	47	5
P_{eq}(W/Hz$^{1/2}$)	2.1×10^{10}	5.9×10^{-11}	1.7×10^{-10}	3.3×10^{-11}
D^*(cm Hz$^{1/2}$/W)	5.0×10^8	4.2×10^8	1.0×10^9	3.6×10^9

Sometimes $D^*/\tau_d^{1/2}$ is used as a figure of merit. For details we refer to Stevens[†] and Kruse et al.[‡]

9.3 THE RESISTIVE BOLOMETER

The resistive bolometer circuit is shown in Fig. 9.2. It consists of a temperature-sensitive resistance R to which a blackened plate of area A is connected. Incident radiation heats this receiving plate and this changes the resistance R. This change in resistance is detected by passing a d.c. current through the resistor and measuring the voltage across it.

We start by writing down the basic equation of the system. The temperature coefficient α of the resistance R is defined as

$$\alpha = \frac{1}{R}\frac{dR}{dT} \qquad (9.27)$$

---

*Strictly speaking, the area A is the *total* area of the receiving plate, namely front plus back.
[†]N. B. Stevens, in R. K. Willardson and A. Beer, (eds.), *Semiconductors and Semimetals*, Vol. 5, Chapter 7, pp. 287–318, 1970.
[‡]P. W. Kruse, L. D. McGlauchlin, and R. B. McQuistan, *Elements of Infrared Technology*, Wiley, New York, 1962.

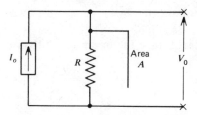

Figure 9.2. Bolometer circuit. A d.c. current generator I_0 is passed through a bolometer of resistance R that is heated by incoming radiation incident on a receiving plate of area A, thus producing an output voltage V_0.

For metal resistors $R = \text{const } T$, and hence

$$\alpha \simeq \frac{1}{T} \tag{9.27a}$$

For extrinsic semiconductors $R = \text{const } T^\gamma$ with $\gamma \simeq \frac{3}{2}$, and hence

$$\alpha \simeq \frac{\gamma}{T} \tag{9.27b}$$

For intrinsic semiconductors of gap width E_g

$$R = \text{const} \exp\left(\frac{eE_g}{2kT}\right)$$

and hence

$$\alpha = -\frac{eE_g}{2kT^2} \tag{9.27c}$$

This is negative and much larger than in the previous cases. For example, for intrinsic germanium ($E_g \simeq 0.70$ V) at $T = 300°$ K we have $\alpha \simeq -0.045$.

The resistance R is heated by the d.c. current I_0 to the temperature T_1 at which its resistance is R_0. If the environment temperature is T_0, then the radiation received is $\eta\sigma A T_0^4$ and the radiation emitted is $\eta\sigma A T_1^4$ (Stefan–Boltzmann law), so that

$$P_0 = I_0^2 R_0 = \frac{V_0^2}{R_0} = \eta\sigma A \left(T_1^4 - T_0^4 \right) \tag{9.28}$$

The heat loss conductance g_H of the detector is given by (9.8a) as

$$g_H = 4\eta\sigma A T_1^3 \tag{9.29}$$

where η is the emissivity of the surface and σ the Stefan–Boltzmann constant, $V_0 = I_0 R_0$ the d.c. voltage developed across the resistance R_0, and P_0 is the electrical power delivered to the resistor. By proper design of the

The Resistive Bolometer

circuit the heat loss due to heat conduction can be made quite small. We shall assume in the subsequent discussions that $\eta = 1$.

9.3a Technical Sensitivity

We now calculate the technical sensitivity by taking I_0 to be a constant, but R varies because of the incident radiation. The equations of the system are, if $\eta = 1$

$$C_H \frac{d\Delta T}{dt} + g_H \Delta T = \Delta(I_0^2 R) + P_1 \exp(j\omega t) \qquad (9.30)$$

$$\Delta V = I_0 \Delta R = I_0 R_0 \alpha \Delta T = \alpha V_0 \Delta T \qquad (9.31)$$

Now

$$\Delta(I_0^2 R) = I_0^2 \Delta R = I_0^2 R_0 \alpha \Delta T = \alpha P_0 \Delta T$$

Substituting into (9.30) yields

$$C_H \frac{d\Delta T}{dt} + (g_H - \alpha P_0)\Delta T = P_1 \exp(j\omega t) \qquad (9.32)$$

Substituting $\Delta T = \Delta T_0 \exp(j\omega t)$ and solving for ΔT_0 yields

$$\Delta T_0 = \frac{P_1}{j\omega C_H + g_H - \alpha P_0} \; ; \quad \Delta V_0 = \frac{\alpha V_0 P_1}{j\omega C_H + g_H - \alpha P_0} \qquad (9.33)$$

Therefore for $\omega^2 C_H^2 \ll (g_H - \alpha P_0)^2$ (low frequencies)

$$\Delta V_0 = \frac{\alpha V_0}{g_H - \alpha P_0} P_1 \qquad (9.33a)$$

which is independent of frequency; for $\omega^2 C_H \gg (g_H - \alpha P_0)^2$ (high frequencies)

$$\Delta V_0 = \frac{\alpha V_0}{\omega C_H} P_1 \qquad (9.33b)$$

which is inversely proportional to ω.

9.3b Device-Impedance Z

Next we evaluate the device-impedance Z. To that end we put a current generator $i_0 \exp(j\omega t)$ in parallel to I_0, calculate the output voltage $v_0 \exp(j\omega t)$, and define Z as v_0/i_0.

We first write down the complete equations

$$C_H \frac{dT}{dt} + \sigma A(T^4 - T_0^4) = I^2 R \quad (9.34)$$

$$V = IR \quad (9.35)$$

where I is the total current and V the total voltage. We next substitute

$$I = I_0 + \Delta I, \quad T = T_1 + \Delta T, \quad V = V_0 + \Delta V$$

to find the small signal equations. We have

$$R = R_0 + \Delta R = R_0 + \alpha R_0 \Delta T$$

Since

$$\Delta(I^2 R) = (I_0 + \Delta I)^2 (R_0 + \alpha R_0 \Delta T) - I_0^2 R_0$$
$$= 2 I_0 R_0 \Delta I + \alpha I_0^2 R_0 \Delta T = 2 V_0 \Delta I + \alpha P_0 \Delta T$$

Substituting into (9.34) yields

$$C_H \frac{d\Delta T}{dt} + (g_H - \alpha P_0) \Delta T = 2 V_0 \Delta I \quad (9.34a)$$

$$\Delta V = I_0 \Delta R + R_0 \Delta I = \alpha V_0 \Delta T + R_0 \Delta I \quad (9.35a)$$

Substituting

$$\Delta T = \Delta T_0 \exp(j\omega t); \quad \Delta I = i_0 \exp(j\omega t); \quad \Delta V = v_0 \exp(j\omega t)$$

and solving for ΔT_0 yields

$$\Delta T_0 = \frac{2 V_0}{j\omega C_H + g_H - \alpha P_0} i_0 \quad (9.36)$$

and, since $V_0^2 = P_0 R_0$,

$$v_0 = \alpha V_0 \Delta T_0 + R_0 i_0 = \left[R_0 + \frac{2\alpha P_0 R_0}{j\omega C_H + g_H - \alpha P_0} \right] i_0$$

so that

$$Z = \frac{v_0}{i_0} = R_0 + \frac{2\alpha P_0 R_0}{j\omega C_H + g_H - \alpha P_0} \quad (9.37)$$

The Resistive Bolometer

For low frequencies $[\omega^2 C_H^2 \ll (g_H - \alpha P_0)^2]$ this is

$$Z = R_0 + \frac{2\alpha P_0 R_0}{g_H - \alpha P_0} = R_0 \frac{g_H + \alpha P_0}{g_H - \alpha P_0} \qquad (9.37a)$$

so that Z is real and larger than R_0 for $\alpha > 0$, due to thermal feedback. At high frequencies, that is, for $\omega^2 C_H^2 \gg (g_H - \alpha P_0)^2$, $Z = R_0$ which corresponds to the d.c. resistance. Note that the device is at the limit of thermal stability at low frequencies if $(g_H - \alpha P_0) = 0$, or if $(g_H + \alpha P_0) = 0$.

9.3c Noise and Equivalent Power P_{eq}

We finally calculate the NEP of the device. Because the emission noise is now at the temperature T_1 and the absorption noise at the temperature T_0, the spectral intensity of the thermal fluctuation noise is for $\eta = 1$

$$S_H(f) = 8\sigma A k T_1^5 + 8\sigma A k T_0^5 \qquad (9.38)$$

The noise equations are

$$C_H \frac{d\Delta T}{dt} + g_H \Delta T = I_0^2 \Delta R + H(t) = \alpha P_0 \Delta T + H(t) \qquad (9.39)$$

or

$$C_H \frac{d\Delta T}{dt} + (g_H - \alpha P_0)\Delta T = H(t) \qquad (9.39a)$$

and

$$\Delta V = I_0 \Delta R + E(t) = \alpha V_0 \Delta T + E(t) \qquad (9.40)$$

where $H(t)$ represents the temperature-fluctuation noise source and $E(t)$, the thermal noise source of the electrical circuit.

Substituting

$$\Delta T = \sum_n a_n \exp(j\omega_n t); \qquad H(t) = \sum_n b_n \exp(j\omega_n t)$$

$$E(t) = \sum_n c_n \exp(j\omega_n t) \qquad \Delta V = \sum_n d_n \exp(j\omega_n t)$$

yields

$$(j\omega_n C_H + g_H - \alpha P_0)a_n = b_n; \qquad a_n = \frac{b_n}{j\omega_n C_H + g_H - \alpha P_0} \qquad (9.41)$$

$$d_n = \alpha V_0 a_n + c_n = \frac{\alpha V_0}{j\omega_n C_H + g_H - \alpha P_0} b_n + c_n \qquad (9.42)$$

Hence

$$S_V(f) = \frac{\alpha^2 V_0^2}{\omega^2 C_H^2 + (g_H - \alpha P_0)^2} S_H(f) + S_E(f)$$

$$= \frac{\alpha^2 P_0 R_0 (8kT_1^5 \sigma A + 8kT_0^5 \sigma A)}{\omega^2 C_H^2 + (g_H - \alpha P_0)^2} + 4kT_1 R_0 \quad (9.43)$$

since $S_E(f) = 4kT_1 R_0$. This equals thermal noise of R_0 for high frequencies $[\omega^2 C_H^2 \gg (g_H - \alpha P_0)^2]$, but at low frequencies (9.43) does not correspond to thermal noise; this is not so surprising, since we have a nonequilibrium situation involving thermal feedback.

We now calculate P_{eq} by putting

$$\frac{\alpha V_0 P_{eq}}{\left[\omega^2 C_H^2 + (g_H - \alpha P_0)^2\right]^{1/2}} = [S_V(f)]^{1/2}$$

which yields

$$P_{eq} = \left[4kT_1 R_0 \frac{\omega^2 C_H^2 + (g_H - \alpha P_0)^2}{\alpha^2 V_0^2} + 8kT_1^5 \sigma A + 8kT_0^5 \sigma A\right]^{1/2} \quad (9.44)$$

At high frequencies $[\omega^2 C_H^2 \gg (g_H - \alpha P_0)^2]$ the first term in (9.44) predominates and, since $(V_0^2/R_0) = P_0$, the equation becomes

$$P_{eq} = \frac{\omega C_H}{|\alpha|} \left(\frac{4kT_1}{P_0}\right)^{1/2} \quad (9.44a)$$

This decreases with increasing P_0, since T_1 increases with increasing P_0, but at a slower rate than P_0.

At low frequencies $[\omega^2 C_H^2 \ll (g_H - \alpha P_0)^2]$, (9.44) becomes

$$P_{eq} = \left[\frac{4kT_1 (g_H - \alpha P_0)^2}{\alpha^2 P_0} + 8kT_1^5 \sigma A + 8kT_0^5 \sigma A\right]^{1/2} \quad (9.44b)$$

We must now consider two cases:

1. α *is positive*, that is, $\alpha = \gamma/T$ with $\gamma = 1$ or $\gamma = \frac{3}{2}$. Then the first term predominates because α^2 is a very small number.

Can $g_H - \alpha P_0$ be zero, so that the device is at the limit of stability? No, for the equation $P_0 = (g_H/\alpha)$ has no solution for T_1. For since $P_0 = \sigma A(T_1^4 - T_0^4)$ and $(g_H/\alpha) = (4\sigma A T_1^4/\gamma)$, the equation may be written

$$\sigma A(T_1^4 - T_0^4) = \frac{4\sigma A T_1^4}{\gamma} \tag{9.44c}$$

and this has no solution for $\gamma < 4$, a value that would not be reached in normal cases. Hence the circuit is thermally stable.

According to Kruse et al.,[†] D^* can have a value of 2×10^8 for metal bolometers of this type.

2. *α is negative.* This gives especially good results in intrinsic material, since α is so large in that case. Since α is negative, the limit of thermal stability is now reached if $(g_H + \alpha P_0) = 0$.

P_0 must now be so chosen that P_{eq} is a minimum. This occurs for a value of T_1 that is not much larger than T_0. We may thus write

$$P_{eq} = \left[4kT_0 \frac{(g_H - \alpha P_0)^2}{\alpha^2 P_0} + 16kT_0^5 \sigma A \right]^{1/2} \tag{9.45}$$

so that we must optimize $(g_H - \alpha P_0)^2/\alpha^2 P_0$. In this case we write

$$P_0 = \sigma A(T_1^4 - T_0^4) = 4\sigma A T_0^3 \Delta T \tag{9.46}$$

where $\Delta T = (T_1 - T_0)$. Hence

$$\frac{(g_H - \alpha P_0)^2}{\alpha^2 P_0} = \frac{4\sigma A T_0^3}{|\alpha|} \frac{(1 + |\alpha|\Delta T)^2}{|\alpha|\Delta T} \tag{9.47}$$

This is a minimum for $|\alpha|\Delta T = 1$ or

$$\Delta T = \frac{1}{|\alpha|} \tag{9.48}$$

and the minimum value is

$$\frac{(g_H - \alpha P_0)^2}{\alpha^2 P_0} = 4\sigma A T_0^3 \cdot \frac{4}{|\alpha|} \tag{9.49}$$

Hence

$$P_{eq} = (16kT_0^5 \sigma A)^{1/2} \left[\frac{4}{|\alpha|T_0} + 1 \right]^{1/2} \tag{9.50}$$

[†]P. W. Kruse, L. D. McGlauchlin, and R. B. McQuistan, *Elements of Infrared Technology*, Wiley, New York, 1962.

where the first term between brackets is the thermal noise contribution. For $\alpha \simeq 0.05$ and $T_0 = 300°$ K this term is small in comparison to unity. In that case P_{eq} is close to the temperature-fluctuation noise limit, unless flicker noise of the current-carrying resistor is a problem.

Unfortunately the condition $|\alpha|\Delta T = 1$ is not physically realizable. For, since α is negative, (9.46) indicates that

$$\alpha P_0 = 4\sigma A T_0^3 \alpha \Delta T = g_H \alpha \Delta T = -g_H, \quad \text{or } g_H + \alpha P_0 = 0.$$

In other words, the device is at the limit of thermal stability; for $|\alpha|\Delta T > 1$ the device burns out.

For stable operation one should thus stay well below the limit $|\alpha|\Delta T = 1$. This is no real handicap, for $\frac{1}{4}(1 + |\alpha|\Delta T)^2/(|\alpha|\Delta T)$ is 9/8 for $|\alpha|\Delta T = \frac{1}{2}$ and 25/16 for $|\alpha|\Delta T = \frac{1}{4}$. Moreover, the second term in (9.45) predominates, so that staying well below the limit of thermal stability does not result in an appreciable deterioration of P_{eq}. It also helps in reducing flicker noise.

REFERENCES

R. C. Jones, *Advances in Electronics*, Vol. V, Adademic, New York, 1953; idem., Vol. XI, ibid., 1959.

M. R. Holter, S. Nudelman, G. H. Suits, W. L. Wolfe, and G. J. Zissis, *Fundamentals of Infrared Technology*, MacMillan, New York, 1962.

10

PHOTOELECTRIC DETECTORS AND CLASSICAL DETECTORS

Photoelectric detectors occur in three forms: (a) photoemissive diodes, (b) photodiodes, and (c) photovoltaic cells. All have in common that the photons of the incident radiation excite carriers, but they differ in what happens to these carriers (Section 10.1).

At high frequencies the device capacitance and the input capacitance of the associated amplifier shunt the output of the device and thus limit the frequency response of the system. Circuits designed to partly overcome this effect are discussed in Section 10.2.

Another way of raising the signal coming out of the device is to use multiplication methods, which are discussed in Sections 10.3 (for photodiodes) and 10.4 (for photoemissive diodes).

Classical detectors are detectors that are not based on quantum effects but that respond *directly* to the square of the applied input voltage. Schottky barrier diodes (either of the extended contact or the point-contact type) and metal–oxide–metal diodes are good examples. By careful design their response can extend into the infrared regime. This problem is discussed in Section 10.5.

To convey information, the incident light beam must be modulated. This can be done by direct modulation of the light source. In some measuring techniques a light chopper is used, converting the d.c. signal into an a.c. signal that can be more easily amplified and more accurately detected with a phase-sensitive technique.

10.1 PHOTOEMISSIVE DIODES, PHOTODIODES, AND PHOTOVOLTAIC CELLS

In the photoemissive diodes the carriers excited by the incident light are electrons; they are excited to a sufficiently high energy inside the photosensitive cathode material so that they can be emitted by the cathode and collected by a positive anode. As a consequence a photocurrent flows; this can be processed by standard electronic techniques.

Photodiodes are junction diodes with the junction located closely to the surface. The incident light generates hole–electron pairs in or near the space-charge region of the p–n junction; if the junction is back-biased, the holes go to the p region and the electrons to the n region so that a photocurrent flows that can be processed by standard electronic techniques.

A photovoltaic cell is a photodiode that is open-circuited for d.c. so that the d.c. current is zero. The hole–electron pairs generated by the incident light then produce an open-circuit voltage, the so-called *photovoltage*, that can be processed by standard electronic techniques.

10.1a Photoemissive Diodes and Photodiodes.

Both types of device can be treated simultaneously. Let us assume that a modulated light beam of power $P_1 \exp(j\omega t)$ is incident upon the device. The a.c. photocurrent is then in either case

$$I_p = \frac{e(1-R)\eta P_1}{h\nu} = (1-R)\eta \frac{P_1}{V_{ph}} \tag{10.1}$$

where η is the quantum efficiency; $V_{ph} = (h\nu/e)$ is the photon energy in eV, and R is the power-reflection coefficient of the surface for the incident radiation. For a photoemissive diode η may be only a few percent and is not much larger than 20% in the best devices; for a good solid-state diode η may be much larger, for instance, $\eta \simeq 0.80 - 0.90$.

To prove (10.1), we observe that $(1-R)P_1$ is the active part of the incident power; the number of quanta used per second is therefore $(1-R)P_1/(h\nu)$, and the number of photoelectrons or hole–electron pairs collected per second is $\eta(1-R)P_1/(h\nu)$. Multiplying by the electron charge e gives the current I_p.

We have for V_{ph}, expressed in terms of the wavelength

$$V_{ph} = \frac{h\nu}{e} = \frac{hc}{\lambda e} = \frac{1.24}{\lambda} \text{ eV} \tag{10.2}$$

where λ is in microns. The quantum energy for 10 μ radiation is therefore 0.124 eV.

For the solid-state diode we must require $V_{ph} > E_g$, where E_g is the gap width of the photodiode material in eV. Values as low as 0.10 eV are obtainable, but in that case one must drastically cool the photodiode to reduce its dark current.

For photoemissive cathodes in which the electron affinity χ is negative, we must also require $V_{ph} > E_g$, since all electrons that arrive in the conduction band can potentially escape. If $\chi > 0$, however, one must require $V_{ph} > E_g + \chi$. The present limit is $\lambda \cong 1.4\ \mu$; in that case the thermionic emission current is quite large and one has to cool the photocathode to improve it. A GaAs—O—Cs emissive cathode is a good example of a photoemissive cathode with negative χ; this value is obtained by a nearly monoatomic O—Cs layer on the surface.

Both types of diode are affected by the dark current I_0 of the device. For the solid-state diode I_0 is simply the back current of the back-biased diode, and for a photocathode it is caused by thermionic emission. In both cases I_0 can be lowered by cooling the device. Assuming that the current I_0 has full shot noise (which is not quite true for the solid-state diode at high frequencies), the noise is given by

$$S_I(f) = 2eI_0 \tag{10.3}$$

Evaluating the NEP P_{eq} by equating

$$(I_p)_{r.m.s.} = (2eI_0)^{1/2}, \quad \text{yields} \quad P_{eq} = \frac{V_{ph}}{(1-R)\eta}(2eI_0)^{1/2} \tag{10.4}$$

D^* is therefore given by

$$D^* = \frac{A^{1/2}}{P_{eq}} = \frac{(1-R)\eta}{V_{ph}} \frac{1}{(2eJ_0)^{1/2}} \tag{10.4a}$$

where J_0 is the dark-current density. For solid-state diodes the concept of D^* must be used with some caution, for I_0 may not be proportional to the device area, so that J_0 is not independent of the device area either (Section 10.1c).

For example in a solid-state photodiode with $R=0$, $\eta = 0.80$, $V_{ph} = 1.0$ V, $I_0 = 10^{-6}$ A, and $P_{eq} = 7.1 \times 10^{-13}$ W/Hz$^{1/2}$, whereas for a photoemissive diode with $R=0$, $\eta = 10^{-2}$, $V_{ph} = 1.0$ V, $I_0 = 10^{-9}$ A, and $P_{eq} = 1.8 \times 10^{-12}$ W/Hz$^{1/2}$. The poorer performance is due to the lower quantum efficiency, despite the smaller dark current.

The signal coming out of a solid-state photodiode or photoemissive diode must be amplified and processed. To that end it is necessary to feed the diode into a load resistance R_L. One must now require that the noise of R_L is small in comparison with the noise of the dark current I_0. This means (Fig. 10.1)

$$2eI_0 > \frac{4kT}{R_L} \quad \text{or} \quad R_L > \frac{2kT}{eI_0} = \frac{1}{20I_0} \tag{10.5}$$

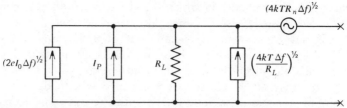

Figure 10.1. Photodiode connected to a load resistance R_L and an amplifier of noise resistance R_n.

at $T = 300°$ K; for $I_0 = 10^{-6}$A, $R_L > 50,000$ Ω. This is easily obtainable, unless R_L must meet certain bandwidth restrictions; that case is discussed in Section 10.2.

At this stage some further discussion is needed about the dark current I_0. For the photoemissive diode

$$I_0 = AT^2 \exp\left(-\frac{e\phi}{kT}\right) S \qquad (10.6)$$

where S is the cathode area, ϕ the work function in eV, and $A = 120$ A cm$^{-2\circ}$ K^{-2}.

For a solid-state photodiode of the p^+–n variety (strongly p-type surface layer)

$$I_0 = e\left(\frac{D_p}{\tau_p}\right)^{1/2} \frac{n_i^2}{N_d} S \qquad (10.7)$$

if recombination occurs in the n region. Here D_p is the hole-diffusion constant, τ_p the hole lifetime in the n region, n_i the intrinsic carrier concentration, and N_d the donor concentration in the n region. For silicon $D_p = 12.5$ cm^2/s, $n_i = 10^{10}$/cm^3; with $\tau_p = 10^{-6}$ s and $N_d = 10^{16}$/cm^3, the dark-current density is 5.7×10^{-12} A/cm^2. Actually, the dark-current density is several orders of magnitude larger in silicon or germanium devices, because most recombination is due to volume recombination in the space-charge region or to surface recombination at the surface of the space-charge region. We come back to that problem in Section 10.1c.

10.1b The Photovoltaic Cell

A photovoltaic cell is an open-circuited photodiode so that the current $I = 0$. Without light the characteristic of the diode near $I = 0$ would be

$$I = I_g(V)\left[\exp\left(\frac{eV}{kT}\right) - 1\right] \qquad (10.8)$$

where $I_g(V)$ is the generation current.* Hence

$$g_0 = \left(\frac{dI}{dV}\right)_0 = \left\{\frac{dI_g}{dV}\left[\exp\left(\frac{eV}{kT}\right) - 1\right] + \frac{eI_g(V)}{kT}\exp\left(\frac{eV}{kT}\right)\right\}_{V=0} = \frac{eI_0'}{kT} \quad (10.8a)$$

where $I_0' = I_g(0)$. Hence the differential resistance is

$$R_0 = \frac{1}{g_0} = \frac{kT}{eI_0'} \quad (10.8b)$$

With light we have

$$I = I_g(V)\exp\left(\frac{eV}{kT}\right) - I_g(V) - I_p = 0; \quad V = \frac{kT}{e}\ln\left[1 + \frac{I_p}{I_g(V)}\right] \quad (10.9)$$

where I_p is the photocurrent. For $I_p \ll I_0'$, $I_g(V) = I_0'$ and

$$V \approx \frac{kT}{e}\frac{I_p}{I_0'} = I_p R_0 \quad (10.9a)$$

Hence in analogy with the diode case

$$P_{eq} = \frac{(4kT/R_0)^{1/2} V_{ph}}{(1-R)\eta} \quad (10.10)$$

where we have simply replaced $2eI_0$ by $(4kT/R_0)$. For $R = 0$, $\eta = 0.80$, $V_{ph} = 1$ V, and $I_0' = 10^{-6}$ A, $R_0 = 25{,}000$ Ω and $P_{eq} = 1.0 \times 10^{-12}$ W/Hz$^{1/2}$. This is $\sqrt{2}$ times as large as in the photodiode case discussed earlier.

From the fact that the noise is due to the flow of two equal and opposite currents, each giving a spectrum $S_i(f)$, we have

$$2S_i(f) = \frac{4kT}{R_0} = 4eI_0', \quad \text{or} \quad S_i(f) = 2eI_0' \quad (10.11)$$

This corresponds to full shot noise of I_0'.

Photovoltaic cells of InSb operate to about 5.5 μ wavelength and CdHgTe and CdSnTe photovoltaic cells of the proper composition operate in the 10 μ wavelength range. Of course, these diodes have to be cooled to at least liquid nitrogen temperature to sufficiently reduce the dark current I_0' and thereby raise the differential resistance R_0 of the device.

---

*This equation follows from (10.12) by substituting $pn = n_i^2 \exp(eV/kT)$ and integrating (10.12) over the space charge region.

10.1c Dark Current in p–n Diodes

We mentioned before that the dark current I_0 in p–n diodes was mainly caused by recombination in the space-charge region. To find I_0 for that case one must turn to the Shockley–Read–Hall recombination theory. According to that theory the net recombination rate R is

$$R = \frac{pn - n_i^2}{(n+n_1)\tau_{p0} + (p+p_1)\tau_{n0}} \quad (10.12)$$

where n_i is the intrinsic carrier concentration, p and n are the hole and electron densities, p_1 and n_1 the hole and electron densities when the Fermi level is at the trapping level, and τ_{p0} and τ_{n0} are lifetimes; τ_{p0} represents the hole lifetime in strongly n-type material, and τ_{n0} represents the electron lifetime in strongly p-type material.

In (10.12) the first term represents the true recombination rate and the second term, the net generation rate. Since the latter is responsible for $I_g(V)$, we have

$$I_g(V) = eS \int_0^d \frac{n_i^2 \, dx}{(n+n_1)\tau_{p0} + (p+p_1)\tau_{n0}} \quad (10.13)$$

where S is the junction area, and d the width of the space-charge region. $I_g(V)$ now depends on the location of the recombination centers and is proportional to the device area S. Moreover $I_g(V)$ does not saturate but increases with increasing back bias.

For strong back bias $n \cong 0$ and $p \cong 0$, so that

$$I_g(V) = I_0 = eS \frac{n_i^2}{n_1 \tau_{p0} + p_1 \tau_{n0}} d \quad (10.13a)$$

But according to diode theory we have for a $p^+ - n$ diode

$$d = \left[\frac{2\varepsilon\varepsilon_0(V_{\text{dif}} - V)}{eN_d} \right]^{1/2} \quad (10.13b)$$

where e is the electron charge, ε is the dielectric constant of the material, $\varepsilon_0 = 8.85 \times 10^{-12}$ F/m, V_{dif} the diffusion potential, V the back bias, and N_d is the donor concentration in the n region. Hence I_0 varies as $(V_{\text{dif}} - V)^{1/2}$ and does not saturate.

The values of I_0, obtained are considerably larger than what would follow from I_0', but still considerably smaller than the values obtained from experiment. Moreover, experimentally one finds I_0 to be proportional to $S^{1/2}$ for a circular junction. Apparently most of the recombination current is proportional to the junction circumference rather than to the junction area S. This comes about because of recombination at the surface of the space-charge region. That gives rise to a current proportional to $2\pi r$, where r is the radius of the junction; since $S = \pi r^2$ the current is indeed proportional to $S^{1/2}$, as found experimentally.

The effect can be overcome by using a guard ring as shown in Fig. 10.2a. Here a circular n^+ region and an annular n^+ region are diffused in; ohmic metalized contacts are made near the rim of the circular ring and at the annular ring, and the surface is passivated by oxidation. If now the same voltage is applied to the contacts, the bulk dark current will flow to the inner contact and the surface dark current to the outer contact, resulting in a large reduction in dark current, as shown in Fig. 10.2b.

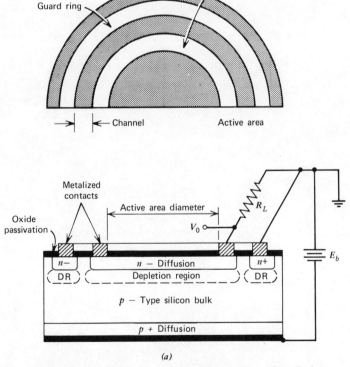

Figure 10.2. (a) Guard ring planar diffused silicon photodiode.

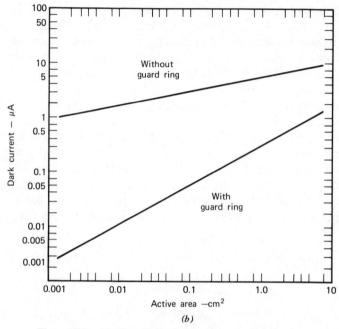

Figure 10.2. (b) Photodiode dark current versus active area.

As an example take an E.E.G.* SHS 100 diode. It has $\eta = 0.85$ at $\lambda = 0.92$ μm. A typical dark current is 10^{-8} A for 5.1 mm² area. Taking (10.4) with $R = 0$ yields

$$P_{eq} = \frac{1.35}{0.85}(2 \times 1.6 \times 10^{-19} \times 10^{-8})^{1/2} = 9.0 \times 10^{-14} \text{ W/Hz}^{1/2}$$

$$D^* = \frac{A^{1/2}}{P_{eq}} = \frac{2.26 \times 10^{-1}}{9.0 \times 10^{-14}} = 2.5 \times 10^{12} \text{ cm Hz}^{1/2}/\text{W}$$

To understand why the high frequency noise of the dark current I_0 is less than full-shot noise consider a recombination center at a distance x from the p^+ region (Fig. 10.3). It alternately emits an electron and a hole. If it emits an electron that is collected by the n region, the transferred charge is $Q_n(x)$; if it emits a hole collected by the p region, the transferred charge is $Q_p(x)$. The net charge transfer due to the generation of a hole–electron pair is $[Q_n(x) + Q_p(x)] = e$, where e is the electron charge. At low frequencies the electrons and holes can be considered fully correlated

---

*E.E.G. stands for Edgerton, Germershausen, and Greer Company.

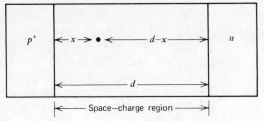

Figure 10.3. $p^+ - n$ Diode with a space-charge region of width d and a generation center at a distance x from the p^+ region.

and the noise is full shot noise. At high frequencies the holes and the electrons can be considered as being generated independently and at random and their effects are evaluated separately. Let $C_0 dx$ be the average rate of generation in the section dx, then

$$I_0 = \int_0^d eC_0 dx = eC_0 d$$

However, the rates $C_0 dx$ of hole and electron generation fluctuate independently and show full shot noise, that is, they have a spectrum $2C_0 dx$. Hence

$$S_I(0) = \int_0^d 2C_0 dx \left[Q_n^2(x) + Q_p^2(x) \right]$$

$$= 2e^2 C_0 \int_0^d \left[1 - 2\frac{Q_n(x)}{e} + 2\frac{Q_n^2(x)}{e^2} \right] dx < 2e^2 C_0 d$$

since $Q_n(x)/e$ lies between 0 and 1. Hence

$$S_I(0) < 2eI_0$$

Lauritzen* has calculated the effect for a uniform field distribution in the space-charge region and finds

$$S_I(0) = \tfrac{2}{3} \cdot 2eI_0$$

Van der Ziel,[†] using a linear field distribution, finds for a p^+-n or n^+-p

---

*P. O. Lauritzen, I.E.E.E. Trans., **ED-15**, 770, 1968.
[†]A. van der Ziel, *Solid State Electronics*, **18**, 969, 1975.

diode

$$S_I(0) = \frac{11}{15} \cdot 2eI_0$$

At zero bias the noise is full shot noise of two equal and opposing currents at low frequencies, as mentioned before. At high frequencies the generated and the recombining electrons and holes can be treated as independent; applying the same reasoning as before, this results in less than full shot noise, accompanied by a corresponding increase in R_0.

10.1d Rise Time

The capacitance C of a photodiode varies as $1/|V|^{1/2}$, where V is the back bias, at least if the junction can be represented as an abrupt junction. The bulk region has a resistance r and the high-frequency network has an equivalent circuit as shown in Fig. 10.4. It is seen that the response time (10–90%) τ of this circuit is

$$\tau = 2.2 Cr \qquad (10.14)$$

Here r is typically of the order of several hundred ohms and C can be made small by using a large back bias. For example, the SHS 100 diode sold by E.E.G. has $C = 4$ pF and $\tau = 4 \times 10^{-9}$ s at 125 V back bias. Hence

$$r = \frac{\tau}{2.2C} = \frac{10^3}{2.2} = 450 \ \Omega$$

To speed up the response, it is therefore important to use a large back bias. In addition it can be advantageous to use a somewhat higher doping in the bulk region, or to use diodes made by an epitaxial technique, to reduce the series resistance r.

10.1e Practical Devices

The most important photoemitters are GaAs (cut-off wavelength 0.91 μm), AgOCs (cut-off wavelength 1.2 μm), Si (cut-off wavelength 1.12 μm), and

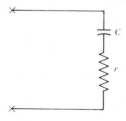

Figure 10.4. Equivalent circuit of a photodiode.

$Ga_{1-x}In_xAs$ (cut-off wavelength 1.4 μm by proper choice of x). The GaAs, Si, and $Ga_{1-x}In_xAs$ must be provided with an —O—Cs monolayer to make the effective electron affinity negative.

The same materials can also be made into p–n junctions and so give photocells or photovoltaic cells. The cut-off wavelengths are the same. In addition Ge p–n junctions (cut-off wavelength 1.76 μm), InSb (cut-off wavelength $\simeq 5.5$ μm) and other compounds are useful. Photovoltaic cells with $Hg_{1-x}Cd_xTe$ and $Pb_{1-x}Sn_xTe$ have become available; with proper choice of x they give good response at 10.6 μm.

Eltek Corporation[†] markets $Hg_{1-x}Cd_xTe$ photovoltaic cells. These have D^* of a few times 10^{10} cm $Hz^{1/2}/W$ at liquid nitrogen temperature; cut-off frequencies f_c as high as several hundred MHz are available for special units. Here f_c is defined by the equation

$$2\pi f_c RC = 1 \tag{10.15}$$

where R is the parallel resistance, and C the parallel capacitance of the junction. Quantum efficiencies as high as 30% or better are possible.[‡]

Photovoltaic cells operating in the 10–12 μm wavelength range are very sensitive to background radiation. To eliminate most of this background radiation the received radiation can be filtered through a material of gap width E'_g close to the photon energy V_{ph} of the received radiation. If E_g is the gap width of the photovoltaic cell, then only the background radiation of photon energy V'_{ph} with $E_g < V'_{ph} < E'_g$ is both passed through the filter and effective in producing hole–electron pairs in the photovoltaic cell. By proper choice of $(E'_g - E_g)$ this amount of radiation can be quite small, and as a consequence the NEP can be reduced considerably.

10.2 BANDWIDTH CONSIDERATIONS

At high frequencies one must take into account the capacitance C of the junction. If the signal is amplified by a JFET amplifier, one must also take into account the input capacitance C_i of the amplifier. The total circuit capacitance is therefore $C + C_i$ and the equivalent circuit is shown in Fig. 10.5.

If R_L is relatively large, then at low frequencies one can neglect the effect of the noise resistance R_n of the amplifier since the noise of load +

[†]Eltek Corporation, Larchmont, N. Y., distributers for the French Company S.A.T. (Societé Anonyme de Telécommunications).
[‡]$PB_{1-x}Sn_xTe$ gives comparable results.

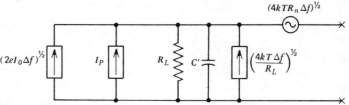

Figure 10.5. Equivalent noise circuit nf a photodiode detector with associated amplifier; $C' = C + C_i$.

amplifier is

$$4kTR_L \Delta f + 4kTR_n \Delta f = 4kTR_L \Delta f \left(1 + \frac{R_n}{R_L}\right)$$

and R_n/R_L is a small number for a good JFET. However at high frequencies the capacitance $C + C_i$ shunts R_L and then the effect of R_n may become significant.

One possible way of improving the situation is to lower R_L to meet the bandwidth requirements. But then the noise of R_L predominates over the noise due to the saturation current I_0. For example, if $C + C_i = 20$ pF and we require an upper cut-off frequency f_1 of 4 MHz (i.e., the response has dropped to 0.707 times the low-frequency value at $f = 4$ MHz), then

$$2\pi f_1 (C + C_i) R_L = 1 \tag{10.16}$$

Substituting the values gives $R_L = 2000$ Ω. Neglecting the noise contribution due to I_0, we have in this case*

$$P_{eq} = \frac{(4kT/R_L)^{1/2} V_{ph}}{(1-R)\eta} \tag{10.17}$$

Substituting $V_{ph} = 1.0$ eV, $R = 0$, $\eta = 0.80$, $T = 300°$ K, and $R_L = 2000$ Ω yields $P_{eq} = 3.6 \times 10^{-12}$ W/Hz. This about 5 times worse than the value calculated in Section 10.1a. The noise in the 4 MHz band thus is $P_n = P_{eq} f_1^{1/2} = 7.2 \times 10^{-9}$ W.

To remedy the situation, one should lower the capacitance C of the diode by applying a larger back bias and designing the diodes such that they can withstand the large back bias. This is not possible for photovoltaic cells, of course. Further improvement is possible by raising R_L to a

---

*We have ignored here the effect of R_n. It can be taken into account by multiplying (10.17) by a factor $[1 + (R_n/R_L)]^{1/2}$.

Bandwidth Considerations

much higher level, thereby spoiling the high-frequency response of the input circuit and then compensating for this frequency response either by applying feedback or by proper design of the frequency response of a subsequent stage. These methods are commonly used in television pickup systems and are discussed in the next section.

10.2a Photodiode Circuit with Improved Noise Performance

The equivalent circuit of the photodiode plus amplifier is shown in Fig. 10.6a; we have put $C + C_i = C'$. This, in turn, is equivalent to the circuit of Fig. 10.6b.* The mean square input noise current at the frequency f is therefore

$$\frac{4kT\Delta f}{R_L} + 2eI_0\Delta f + \frac{4kTR_n\Delta f}{R_L^2} + 4kTR_n\Delta f \omega^2 C'^2 \qquad (10.18)$$

Integrating this over the passband $0 \leqslant f \leqslant f_1$ gives an r.m.s. input-current generator

$$\left(\overline{i^2}\right)^{1/2} = (4kTf_1)^{1/2} \left[\frac{1}{R_L} + \frac{eI_0}{2kT} + \frac{1}{3}R_n(2\pi f_1 C')^2\right]^{1/2} \qquad (10.19)$$

where we have neglected the term R_n/R_L^2 between brackets. Now if $T = 300°$ K, $f_1 = 4.0$ MHz, $R_L = 50{,}000$ Ω, $I_0 = 1$ μA, $R_n = 100$ Ω, and $C' = 20$ pF, we have $1/R_L = 2.0 \times 10^{-5}$ mho, $(eI_0/2kT) = 2.0 \times 10^{-5}$ mho, and $\frac{1}{3}R_n(2\pi f_1 C')^2 = 0.8 \times 10^{-5}$ mho, so that the effect of R_L and I_0 predominates here.

We now define the equivalent noise power P_n by

$$\frac{(1-R)\eta}{V_{ph}} P_n = \left(\overline{i^2}\right)^{1/2} \qquad (10.20)$$

Substituting $R = 0$, $\eta = 0.80$ and $V_{ph} = 1.0$ eV yields

$$P_n = \frac{\left(\overline{i^2}\right)^{1/2}}{0.80} = 2.2 \times 10^{-9} \text{ W}$$

This is about a factor 3 better than in the case of a low-impedance input circuit.

---
*We have neglected here the relatively small contribution of the high-frequency gate noise of the FET.

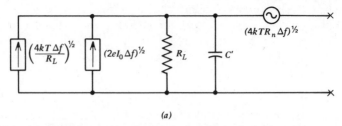

Figure 10.6. (a) Equivalent-noise circuit of photodiode, including effect of the circuit capacitance C'.

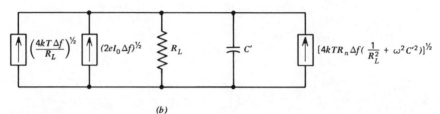

Figure 10.6. (b) Alternate equivalent circuit, in which the noise emf $(4kTR_n\Delta f)^{1/2}$ is replaced by an equivalent-current generator.

By cooling the detector, the term $eI_0/2kT$ is reduced to a negligible value and this lowers P_n by 30%. This may not be worth the effort.

10.2b Photovoltaic Cell Circuit with Improved Noise Performance

In a photovoltaic circuit we can eliminate R_L altogether and compensate for the poor frequency response in the input circuit, either by feedback or by compensation in a subsequent stage. The equivalent circuit corresponding to Fig. 10.6b is then as shown in Fig. 10.7. Hence the corresponding r.m.s. input-current generator is

$$\left(\overline{i^2}\right)^{1/2} = (4kTf_1)^{1/2}\left[\frac{1}{R_0} + \frac{1}{3}R_n(2\pi f_1 C')^2\right]^{1/2} \qquad (10.21)$$

where we have neglected the term R_n/R_0^2 between brackets. Now if $C' = 20$ pF, $R_n = 100$ Ω, $f_1 = 4$ MHz, $I_0 = 10^{-6}$ A, then $1/R_0 = 4 \times 10^{-5}$ mho, $\frac{1}{3}R_n(2\pi f_1 C')^2 = 0.8 \times 10^{-5}$ mho. Calculating P_n for $R=0$, $\eta = 0.80$ and $V_{ph} = 1.0$ eV yields $P_n = 2.2 \times 10^{-9}$ W, as in the previous case. By cooling one can eliminate the term $1/R_0$ and reduce P_n by a factor $(6)^{1/2} = 2.45$, which may be worthwhile.

Multiplication Methods in Junction Diodes

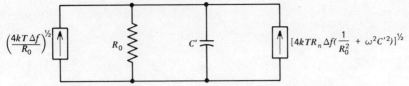

Figure 10.7. Equivalent noise circuit of a photovoltaic cell.

The example given here is probably somewhat too flattering for the photovoltaic cell. For the capacitance C' is much larger than in the photodiode case, because the capacitance C of a junction diode is much larger for zero bias than for a large back bias.

10.3 MULTIPLICATION METHODS IN JUNCTION DIODES

We saw that the a.c. signal in solid-state photodiodes and photovoltaic cells is often quite small, especially if the incident power is small. In that case the amplifier noise may be quite important. It would now be very convenient if the signal could be amplified internally before being processed by an amplifier. We discuss here two such amplification methods.

10.3a The Phototransistor*

Here a transistor with floating base provides the required amplification. This operation can be understood as follows. The Ebers–Moll equations for an n–p–n transistor with a forward-biased emitter and a back-biased collector may be written as (Fig. 10.8)

$$I_E = -I_{Es}\left[\exp\left(\frac{eV_{BE}}{kT}\right) - 1\right] - \alpha_R I_{Cs} \quad (10.22)$$

$$I_C = \alpha_F I_{Es}\left[\exp\left(\frac{eV_{BE}}{kT}\right) - 1\right] + I_{Cs} \quad (10.23)$$

where I_{Es} and I_{Cs} are saturation currents, α_F and α_R are the forward and reverse current amplification factors, and V_{BE} is the base-emitter voltage. Expressing I_C in terms of I_E yields

$$I_C = -\alpha_F I_E + I_{C0}; \quad I_{C0} = (1 - \alpha_R \alpha_F)I_{Cs} \quad (10.24)$$

where I_{C0} is called the *collector saturated current*.

*F. de la Moneda et al., *I.E.E.E. Trans.*, **ED-18**, 340 (1971).

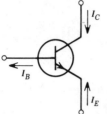

Figure 10.8. Current flow in an *n-p-n* transistor.

Without light the base current is therefore (see Fig. 10.8)

$$I_B = I_C + I_E = (1 - \alpha_F)I_E + I_{C0} \tag{10.25}$$

With the light the base current is

$$I_B = (1 - \alpha_F)I_E + I_{C0} + I_{ph}; \quad I_{ph} = \frac{(1-R)\eta P_1}{V_{ph}} \tag{10.26}$$

where I_{ph} is the photocurrent, R the reflection coefficient for the incident light, η the quantum efficiency, V_{ph} the photon energy, and P_1 the incident (modulated) power. However since the base is floating, I_B is identically zero and hence

$$-I_E = I_C = \frac{I_{C0}}{1 - \alpha_F} + \frac{I_{ph}}{1 - \alpha_F} \simeq h_{FE}I_{C0} + h_{FE}I_{ph} = I_{CE0} + h_{FE}I_{ph} \tag{10.27}$$

where $I_{CE0} = h_{FE}I_{C0}$ and $h_{FE} = \alpha_F/(1 - \alpha_F)$ is the d.c. current amplification factor for the common circuit. Here it is assumed that $h_{FE} \gg 1$; in a good phototransistor $h_{FE} > 100$.

At high frequencies the a.c. signal becomes

$$\frac{h_{fe}}{1 + jf/f_\beta} I_{ph} \tag{10.28}$$

where h_{fe} is the a.c. current amplification factor and f_β is the β cut-off frequency of the transistor; h_{fe} is usually somewhat larger than h_{FE} for reasons that will not be explained here.

Now about the noise. The base current for floating base consists of two equal and opposite currents I_{C0}, each showing full-shot noise. Hence the base-current noise must be written

$$\overline{i_b^2} = 4eI_{C0}\Delta f \tag{10.29}$$

Multiplied by the factor

$$\frac{h_{fe}^2}{1+f^2/f_\beta^2} \tag{10.30}$$

we obtain the contribution of $\overline{i_b^2}$ to the collector noise $\overline{i_c^2}$. In addition the collector current $I_{CE0}=h_{FE}I_{C0}$ shows always full-shot noise. Hence

$$\overline{i_c^2} = 4eI_{C0}\frac{h_{fe}^2}{1+f^2/f_\beta^2}\Delta f+2eh_{FE}I_{C0}\Delta f \tag{10.31}$$

We now evaluate P_{eq} by equating

$$\frac{(1-R)\eta}{V_{ph}}P_{eq}\frac{h_{fe}}{\left(1+f^2/f_\beta^2\right)^{1/2}} = \left(\frac{\overline{i_c^2}}{\Delta f}\right)^{1/2}$$

and obtain

$$P_{eq}= \frac{V_{ph}}{(1-R)\eta}\left[4eI_{C0}+2eI_{C0}h_{FE}\frac{\left(1+f^2/f_\beta^2\right)}{h_{fe}^2}\right]^{1/2} \tag{10.32}$$

For low frequencies, since $h_{FE}/h_{fe}^2 \ll 1$,

$$P_{eq}= \frac{V_{ph}}{(1-R)\eta}(4eI_{C0})^{1/2} \tag{10.33}$$

which is just what would be expected if the base were used as the output electrode of a photovoltaic cell. *Hence an amplification by a factor h_{fe} is obtained that is practically noiseless.*

At high frequencies the last term in (10.32) begins to predominate and the noise of I_{CE0} becomes important. We may thus define an upper cut-off frequency f_1 for which the noise has just doubled. Then

$$4eI_{C0}=2eI_{C0}h_{FE}\frac{\left(1+f_1^2/f_\beta^2\right)}{h_{fe}^2}, \quad \text{or} \quad f_1 \simeq f_\beta\left(\frac{2h_{fe}^2}{h_{FE}}\right)^{1/2} \tag{10.34}$$

Taking $f_\beta=20$ kHz (the currents are so small!) and $h_{fe}=h_{FE}=100$ yields $f_1=300$ kHz. This is good enough for the audio but not for the video range of signals; the frequency response can be extended by making f_β larger.

10.3b Avalanche Multiplication

Another way of multiplying the signal generated in a diode is to use avalanche multiplication. Avalanche diodes can be built that show uniform breakdown and uniform current multiplication before breakdown. We shall see how this results in an improvement in the signal-to-noise ratio.

The signal calculation is quite simple if the electrons and holes have equal ionizing power. Let I_0 be the diode dark current before multiplication and let the photocurrent $I_{ph} \ll I_0$. If p hole–electron pairs are generated when an individual hole–electron pair traverses the space-charge region once, where $p<1$, then the multiplied photocurrent is

$$I_{ph}(1+p+p^2+\cdots)=\frac{I_{ph}}{1-p}=MI_{ph} \tag{10.35}$$

where $M=1/(1-p)$ is the current multiplication factor. By making p close to unity, a large multiplication can be obtained.

Now the noise. The spectral intensity of the primary current is $2eI_0$. Because the hole–electron pairs are generated independently and at random, the generated current due to a single traversing of the junction by hole–electron pairs shows full-shot noise; that is, the currents $pI_0, p^2I_0, \ldots$ all show full-shot noise. All these noise contributions must now be added and multiplied by M^2. Hence

$$S_I(f)=2e(I_0+pI_0+p^2I_0\cdots)M^2=\frac{2eI_0}{1-p}M^2=2eI_0M^3 \tag{10.36}$$

McIntyre* has calculated the noise for the case that electrons and holes have different ionization coefficients α and β, respectively. It is found that the ratio $k=(\beta/\alpha)$ is practically independent of the applied voltage. Expressed in terms of k and M, McIntyre obtained the following:

1. p^+-n diode (all current carried by holes)

$$S_I(f)=2eI_0M^3\left[1+\left(\frac{1-k}{k}\right)\left(\frac{M-1}{M}\right)^2\right] \tag{10.37}$$

2. n^+-p diode (all current carried by electrons)

$$S_I(f)=2eI_0M^3\left[1-(1-k)\left(\frac{M-1}{M}\right)^2\right] \tag{10.38}$$

The first expression is quite large for $k\ll 1$ and quite small for $k\gg 1$, so that the lowest noise is obtained if the holes have the largest ionizing

*R. J. McIntyre, *I.E.E.E. Trans., Electron Devices*, **ED-13**, 164 (1966).

power. The second expression is quite large when $k \gg 1$ and quite small for $k \ll 1$, so that the lowest noise is obtained if the electrons have the largest ionization coefficient. These predictions have been well verified by experiment. In germanium the holes have the largest ionizing power $(k>1)$, whereas in silicon the electrons have by far the largest ionizing power. For GaAs and GaP $k \simeq 1$.

At first sight it would seem that multiplication has an adverse effect since the signal is multiplied by M, but the noise by $M^{3/2}$. However the amplifier noise usually predominates by far, and then a significant improvement in signal-to-noise ratio is possible.

To demonstrate this, we consider that the amplifier has been so designed that it has a very high impedance of the input circuit plus frequency compensation by feedback or in a subsequent amplifier stage to achieve an upper cut-off frequency f_1. The noise can then be represented by a current generator $(2eI_{eq}f_1)^{1/2}$ in parallel with the amplifier input. If the amplifier noise predominates, we have (according to Section 10.2)

$$2eI_{eq}f_1 = 4kTf_1 \cdot \frac{1}{3}R_n(2\pi f_1 C')^2, \quad \text{or} \quad I_{eq} = \frac{2kT}{e} \cdot \frac{1}{3}R_n(2\pi f_1 C')^2 \quad (10.39)$$

Using $f_1 = 4$ MHz, $T = 300°$ K, $R_n = 100$ Ω, $C' = 20$ pF, and $R_L = \infty$ yields $I_{eq} = 0.4$ μA. Since a good silicon diode with guard ring has $I_0 = 10^{-9}$ A, the noise of the amplifier predominates by far.

Without multiplication we thus have for the signal-to-noise power ratio

$$\frac{S}{N} = \frac{I_{ph}^2}{2e(I_0 + I_{eq})f_1} = \frac{I_{ph}^2}{2eI_0f_1(1 + I_{eq}/I_0)} \quad (10.40)$$

where I_{ph} now represents the r.m.s. value of the photocurrent and $I_{eq} \gg I_0$. *With* multiplication we have

$$\frac{S}{N} = \frac{M^2 I_{ph}^2}{2e(I_0 M^3 + I_{eq})f_1} = \frac{I_{ph}^2}{2eI_0f_1[M + (I_{eq}/I_0)/M^2]} \quad (10.41)$$

This has a minimum value if $M = (2I_{eq}/I_0)^{1/3}$ and this minimum value is

$$\frac{S}{N} = \frac{I_{ph}^2}{2eI_0f_1}\left[\frac{2}{3}\left(\frac{I_0}{2I_{eq}}\right)^{1/3}\right] \quad (10.41a)$$

This is much better than without amplification. If $I_{eq} = 0.4$ μA and $I_0 = 10^{-9}$ μA

$$1 + \frac{I_{eq}}{I_0} \cong 400 \quad \text{and} \quad \frac{3}{2}\left(\frac{2I_{eq}}{I_0}\right)^{1/3} \cong 14$$

so that an improvement in signal-to-noise power ratio by a factor 29 has been obtained. The value of the equivalent noise power P_n has been improved by the same factor.

10.4 MULTIPLICATION IN PHOTOEMISSIVE DEVICES

The advantage of photoemissive devices is that the emitted electrons move in a vacuum. They can thus be multiplied by secondary emission multiplication and overcome the amplifier noise of the system processing the photoelectric signals. The photomultiplier uses discrete dynodes, and the channel multiplier uses distributed multiplication.

10.4a The Photomultiplier

According to (4.24) the spectral intensity of the noise coming out of a dynode multiplying a primary current I_{pr} by a factor δ is

$$S_I(f) = 2eI_{pr}\delta^2 + 2eI_{pr}\delta(\kappa - \delta) \tag{10.42}$$

where the first term is the multiplied primary noise and the second term is the secondary emission noise. The noise parameter κ is often nearly equal to $\delta + 1$ so that $\kappa - \delta \simeq 1$.

We now put n dynodes with the same parameters κ and δ one behind the other. The amplification is then δ^n and the output noise is the multiplied noise of the primary current plus the amplified secondary emission noise of each dynode, that is,

$$S_i(f) = 2eI_{pr}\delta^{2n} + 2eI_{pr}[\delta(\kappa-\delta)]\delta^{2n-2}$$
$$+ 2eI_{pr}\delta[\delta(\kappa-\delta)]\delta^{2n-4} + \ldots + 2eI_{pr}\delta^{n-1}[\delta(\kappa-\delta)]$$
$$= 2eI_{pr}\delta^{2n}\left[1 + \frac{\kappa-\delta}{\delta}\left(1 + \frac{1}{\delta} + \frac{1}{\delta^2} + \ldots + \frac{1}{\delta^{n-1}}\right)\right]$$
$$\simeq 2eI_{pr}\delta^{2n}\left[1 + \frac{\kappa-\delta}{\delta(1-1/\delta)}\right] = \left(\frac{\kappa-1}{\delta-1}\right)2eI_{pr}\delta^{2n} \tag{10.43}$$

The factor $(\kappa-1)/(\delta-1)$ is called the noise-deterioration factor Γ of the multiplier

$$\Gamma = \frac{\kappa-1}{\delta-1} \simeq \frac{\delta}{\delta-1} \tag{10.43a}$$

Consequently, $\delta = 4$ gives $\Gamma \simeq 4/3$, so that the multiplier gives practically noiseless amplification.

10.4b The Channel Multiplier

The channel multiplier consists of a thin, slightly conducting, hollow tube, the inside of which can emit secondary electrons. If a primary electron beam is now focused into the multiplier and a voltage of 1500–2000 V is applied between the end points of the tube, the secondary electrons are multiplied by *distributed* multiplication that can easily amount to a total multiplication of 10^6 or more.

To describe this distributed multiplication effect, we introduce the multiplication factor p per unit length and write

$$dI(x) = I(x) p \, dx; \quad \frac{dI(x)}{I(x)} = d\ln[I(x)] = p \, dx$$

$$\frac{I(x)}{I_{\text{pr}}} = \exp(px); \quad G = \frac{I(L)}{I_{\text{pr}}} = \exp(pL) \quad (10.44)$$

where $dI(x)$ is the electron current generated in the section dx and G is the current gain of the channel multiplier.

Now we investigate the noise. Since the primary current I_{pr} and the current $dI(x)$ both give full-shot noise, and the noise at x is multiplied by $\exp[p(L-x)]$,

$$S_I(f) = 2eI_{\text{pr}} \exp(2pL) + \int_0^L 2e[I_{\text{pr}} \exp(px) d(px)] \exp[2p(L-x)]$$

$$= 2eI_{\text{pr}} \exp(2pL) \left[1 + \int_0^L \exp(-px) d(px) \right]$$

$$= 2eI_{\text{pr}} \exp(2pL)[2 - \exp(-pL)] = 2eI_{\text{pr}} \exp(2pL)\Gamma \quad (10.45)$$

so that the noise-deterioration factor $\Gamma = 2$ for large gain.

Measurements by Timm* roughly agree with this prediction. However, there was one complication in that the gain G depended on the primary current, at least for the large primary currents used. The gain G is then not I_s/I_{pr}, where I_s is the secondary current, but $G = (\partial I_s / \partial I_{\text{pr}})$. We refer to the paper for details.

Γ can be made smaller than 2.0 by giving the input end of the channel a much higher δ, and by letting the (photo)electrons impinge on that input

*G. W. Timm and A. van der Ziel, *I.E.E.E. Trans.*, **ED-15**, 314 (1968).

with a higher energy. In that case most of the noise comes from this first multiplication, and $\Gamma \simeq (\delta+2)/\delta$, as follows from repeating the derivation of (10.45).

10.5 CLASSICAL DETECTORS

Classical detectors are detectors based on the principle that they respond directly to the square of the a.c. input voltage. Schottky barrier diodes and metal—oxide—metal diodes are examples.

The principle is as follows. Let $I = I(V)$ be the characteristic of the device, let V_0 be the quiescent voltage, and let I_d be the quiescent current. Then, if $V = (V_0 + \Delta V)$, the new current is

$$I = I_d + \left.\frac{dI}{dV}\right|_{V_0} \Delta V + \frac{1}{2} \left.\frac{d^2I}{dV^2}\right|_{V_0} \Delta V^2 + \cdots \tag{10.46}$$

so that the device presents an input conductance

$$g = \left(\frac{dI}{dV}\right)_{V_0} \tag{10.47}$$

to the change ΔV in voltage. If $\Delta V = v_0 \cos \omega t$, the change in d.c. current is therefore

$$\Delta I_d = \frac{1}{2} \left.\frac{d^2I}{dV^2}\right|_{V_0} v_0^2 \langle \cos^2 \omega t \rangle = \frac{1}{4} \left.\frac{d^2I}{dV^2}\right|_{V_0} v_0^2 \tag{10.48}$$

where $\langle \rangle$ denotes an average over a complete cycle, so that ΔI_d is proportional to the square of the a.c. amplitude v_0, as stated.

If the signal source is matched to the detector, the available power P_a of the signal source is dissipated by the detector, so that $P_a = \frac{1}{2} v_0^2 g_0$, or $v_0^2 = (2P_a/g_0)$. Hence

$$\Delta I_d = \frac{1}{2} \frac{(d^2I/dV^2)_{V_0}}{(dI/dV)_{V_0}} P_a \tag{10.49}$$

The sensitivity of the detector to incident power is thus given by the factor

$$\frac{(d^2I/dV^2)_{V_0}}{(dI/dV)_{V_0}} \tag{10.49a}$$

This factor should thus be made as large as possible.

Classical Detectors

To find the NEP, P_{eq}, of the detector, we equate ΔI_d to the square root of the spectral intensity $S_I(f)$ of the short-circuit current of the device under the existing operating conditions. This yields

$$P_{eq} = 2 \frac{(dI/dV)_{V_0}}{(d^2I/dV^2)_{V_0}} [S_I(f)]^{1/2} \qquad (10.50)$$

For zero bias the noise is thermal noise of the conductance g, for nonzero bias the existing devices give shot noise of the current I_d. That is,

$$S_I(f) = 4kT \left. \frac{dI}{dV} \right|_{V_0} \qquad \text{for} \quad V = 0 \qquad (10.51)$$

and

$$S_I(f) \cong 2eI_d \qquad \text{for} \quad V \neq 0 \qquad (10.51a)$$

Section 10.5a applies these considerations to Schottky barrier diodes and Section 10.5b applies them to metal—oxide—metal diodes. The advantage of these devices is that they do not need cooling, such in contrast with the photovoltaic cells operating at 10 μm wavelength.

10.5a Schottky Barrier Diodes

A Schottky barrier diode consists of a rectifying metal contact made on a piece of semiconductor; the rectifying properties are obtained because the metal and the semiconductor are so chosen that a physical barrier is formed between the metal and the semiconductor that impedes the flow of current between metal and semiconductor.

The diode has a characteristic

$$I = I_0 \left[\exp\left(\frac{eV}{kT}\right) - 1 \right] \qquad (10.52)$$

so that

$$g_0 = \left. \frac{dI}{dV} \right|_{V_0} = \frac{e(I + I_0)}{kT} ; \qquad \left. \frac{d^2I}{dV^2} \right|_{V_0} = \frac{e^2(I + I_0)}{(kT)^2}$$

Hence

$$P_{eq} = \frac{2kT}{e} (4eI_0)^{1/2} \qquad (10.52a)$$

for zero bias and

$$P_{eq} = \frac{2kT}{e}(2eI_d)^{1/2} \tag{10.52b}$$

for large forward bias $(I_d \gg I_0)$. The larger current is sometimes needed for an easier matching to the signal source.

We must now apply these considerations to point-contact Schottky barrier diodes operating at infrared frequencies. One must then take into account the capacitance C of the space-charge region, which is in parallel to the high-frequency conductance g_i of the space charge region; g_i may differ from the low-frequency conductance $g_0 = (dI/dV)_{V_0}$ at the operating point, because of electron transit-time effects. Finally, one must take into account that the rectification deteriorates at the highest frequencies because of electron-transit time effects in the space-charge region.

Figure 10.9a shows a point-contact diode, consisting of a metal cat's whisker with a very fine point, pressed against a highly doped semiconductor chip. The part of the cat's whisker between the bend in the whisker and the semiconductor (length L) acts as an antenna with a radiation resistance R_a of about 100 Ω. The equivalent circuit is shown in Fig. 10.9b, $v_a \exp(j\omega t)$ is the antenna e.m.f., $v_0 \exp(j\omega t)$ the voltage developed across the space charge region, the bulk region* is represented by its series resistance r, and the space-charge region by its C, g_i parallel combination. As is well known,

$$r = \frac{\rho}{4a}; \quad C = \frac{\varepsilon \varepsilon_0 A}{d}; \quad d = \left[\frac{2\varepsilon \varepsilon_0}{eN_d}(V_{\text{dif}} - V_0)\right]^{1/2} \tag{10.53}$$

where ρ is the resistivity of the bulk material, a the contact radius, A the contact area, d the width of the space-charge region, ε the dielectric constant of the semiconductor material, and $V_{\text{dif}} - V_0$ the barrier height.

If we now solve the circuit problem of Fig. 10.9b, we obtain

$$v_0 = \frac{v_a}{1 + (g_i + j\omega C)(R_a + r)}$$

Since $P_a = |v_a|^2/8R_a$ is the available antenna power, we write

$$|v_0|^2 = \frac{8P_a R_a}{[1 + g_i(R_a + r)]^2 + \omega^2 C^2 (R_a + r)^2} \tag{10.54}$$

---

*It would be more realistic to represent the bulk region by the resistance r and an inductance L in series, in parallel to the bulk capacitance C_b. The bulk region can then show resonance effects. The inductance is caused by carrier collisions with the lattice.

Classical Detectors

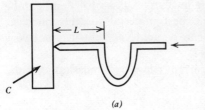

Figure 10.9. (*a*) Point contact semiconductor Schottky barrier diode with a cat's whisker having a sharp point pressed against a semiconductor chip *C*. In the metal—oxide—metal diode the chip is replaced by a metal plate covered by an oxide layer 10–20 Å thick.

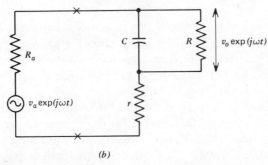

Figure 10.9. (*b*) Equivalent circuit of a point-contact Schottky barrier diode; $R=1/g_i$. The cat's whisker acts as an antenna of radiation resistance R_a ($R_a \simeq 100\ \Omega$).

The transit-time effect on the rectification can be represented by a transit-time deterioration factor $|g(j\omega)|^2$, which is unity at low frequencies and decreases with increasing frequency at the highest frequencies.* We obtain instead of (10.49)

$$\Delta I_d = \frac{1}{2}\left(\frac{d^2I/dV^2}{dI/dV}\right)_{V_0} P_a \frac{4g_0 R_a |g(j\omega)|^2}{[1+g_i(R_a+r)]^2 + \omega^2 C^2 (R_a+r)^2} \quad (10.55)$$

and hence (10.50) may be written

$$P_{eq} = 2\left(\frac{dI/dV}{d^2I/dV^2}\right)_{V_0} (2eI_{d0})^{1/2} \frac{[1+g_i(R_a+r)]^2 + \omega^2 C^2 (R_a+r)^2}{4g_0 R_a |g(j\omega)|^2} \quad (10.56)$$

where I_{d0} is the quiescent current.

We now observe that C, g_i, g_0, and I_{d0} are proportional to the contact area A and that r is proportional to $A^{-1/2}$, so that the best results are obtained for relatively small areas A. Since this would increase the effect of the series resistance r, the resistivity of the bulk material should be made as

*A. van der Ziel, *Physica* 81 *B&C*, 107, 1976.

small as possible by high doping. Doping levels as high as $10^{19}/\mathrm{cm}^3$ can be obtained in n-type GaAs, which is the best material for this purpose. Contact areas A as small as $10^{-11}\mathrm{cm}^2$ can be achieved, but it is not very worthwhile to make A so small that $r > R_a$.

A calculation* shows that when the current flows by "emission over the barrier," then $|g(j\omega)|^2 = 0.25$ at

$$f = \frac{3}{2\pi}\left(\frac{e^2 N_d}{m^* \varepsilon \varepsilon_0}\right)^{1/2} \qquad (10.56a)$$

where e is the electron charge, N_d the donor concentration, and m^* is the effective electron mass. Substituting numbers for GaAs, which has $\varepsilon = 11$ and $m^* = 0.068\, m$, where m is the free-electron mass, one finds that the frequency given by (10.56a) corresponds to a wavelength of 3.0 μm for $N_d = 10^{19}/\mathrm{cm}^3$.

At these high doping levels the current flow is by tunneling through the space-charge region, and that makes $|g(j\omega)|^2$ closer to unity. Practically normal rectification should thus be possible at 10 μm wavelength. Heavily doped point-contact n-type GaAs diodes with small contact area A thus look most promising as detectors in that wavelength range.

10.5b The Metal—Oxide—Metal Diode

The metal—oxide—metal diode consists of a metal plate with a thin oxide layer, about 10 Å (10^{-7} cm) thick, to which a cat's whisker with a very fine point is pressed. The geometric arrangement is as shown in Fig. 10.9a, with the semiconductor chip replaced by the metal plate, and the equivalent circuit is shown in Fig. 10.10; here R_a is the radiation resistance of the antenna and the detector is replaced by an RC parallel combination, with $1/R = (dI/dV)_{V_0}$.

At the highest frequencies the capacitance C shunts the resistance R; the contact area is therefore chosen so that $\omega C R_a < 1$ for the operating frequency. Since $C = (\varepsilon \varepsilon_0 A/d)$, where ε is the dielectric constant of the oxide, A the contact area and d is the oxide thickness, it follows that $A \simeq (200\ \text{Å})^2$ at 10 μm. This is achievable with present techniques. In that case $v_d \simeq v_a$.

If the metal plate and the metal contact are made of identical material, the device has an antisymmetric characteristic and hence no rectification occurs at zero bias. For a forward-biased diode, however, the rectification is quite significant. Another drawback of the device is that, because of the small area, the device has a large amount of flicker noise, since at a given current flicker noise is inversely proportional to the device area.

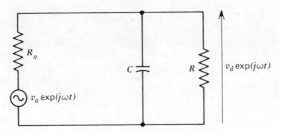

Figure 10.10. Equivalent circuit of a metal—oxide—metal diode.

10.5c Using Classical Detectors to Make Beats Between Optical Signals of Different Frequencies

Schottky barrier diodes and metal—oxide—metal diodes find widespread use in obtaining beats between optical signals of different frequencies. In this way one can extend frequency measurements into the optical range. The principle is that, because of the nonlinear characteristic of these devices, two large signals of frequencies ω_1 and ω_2 produce beat frequencies $\omega_3 = |n_1 \omega_1 - n_2 \omega_2|$. In particular, if $n_1 = 1$, $n_2 = 10$, and $\omega_1 \cong 10 \omega_2$, a low-frequency beat can be obtained, whose frequency can either be measured by standard techniques or can be made to beat with a signal of still lower frequency ω_4. If ω_3 and ω_2 are known, ω_1 can be determined.

In this manner beats between lower-frequency signals and laser signals of 5 or 10 μm wavelength have been obtained. We refer to papers by Dr. Javan's group at the Massachusetts Institute of Technology for details.*

REFERENCES

R. K. Willardson and A. C. Beer, (eds.), *Semiconductors and Semimetals*, Vol. 5, 1970; see especially Chapters 2–5.

---

*J. M. Small, G. M. Elchinger, A. Javan, A. Sanchez, F. J. Bachner, and D. L. Smythe, *Appl. Phys. Lett.*, **24**, 275 (1974).

11

PHOTOCONDUCTIVE DETECTORS

In a photoconductive detector the incident quanta create hole–electron pairs if the photon energy $V_{ph} = (h\nu/e)$ is larger than the gap width E_g between the valence band and the conduction band. These hole–electron pairs increase the conductivity of the semiconductor material used as a photoconductor and this increase in conductivity is detected by applying a d.c. voltage V_0 to the detector D and taking the signal off from a load resistor R. Figure 11.1a shows the experimental arrangement, and Fig. 11.1b shows a circuit used in conjunction with a photoconductive detector.

Section 11.1 discusses the photoconductive response, the noise response, and the maximum signal transfer. Section 11.2 discusses various types of photoconductors—two with band-to-band excitation with different types of recombination, and two with excitation from donor or acceptor levels with or without counterdoping. Section 11.3 discusses the effect of thermal noise, amplifier noise, and flicker noise. Practical circuits are dealt with in Section 11.4. Practical examples of photoconductors are discussed in Section 11.5.

11.1 PHOTOCONDUCTIVE RESPONSE

11.1a Signal Response

Let P be the incident radiation of frequency ν, or of photon energy $V_{ph} = (h\nu/e)$. Then the rate of hole–electron pair production in the photoconductor is

$$Q = \frac{\eta P(1-R)}{eV_{ph}} \tag{11.1}$$

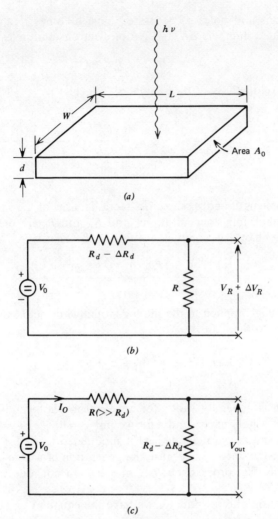

Figure 11.1. Photoconductive detector: (*a*) quanta $h\nu$ are incident on a semiconductor chip of receiving area $A = wL$ and cross-sectional area $A_0 = wd$, where L, w, and d are the dimensions of the semiconductor chip; (*b*) one possible circuit arrangement, in which the voltage is taken from a load resistance R; (*c*) alternate circuit arrangement, in which the voltage is taken from the photoconductor.

If the electrons and holes so generated have lifetimes τ_n and τ_p, respectively, the added numbers ΔN and ΔP of electrons and holes are

$$\Delta N = Q\tau_n; \qquad \Delta P = Q\tau_p \tag{11.2}$$

The dark conductivity of the sample is (Fig. 11.1a)

$$g_0 = e(\mu_n n_0 + \mu_p p_0)\frac{A_0}{L} = e\left[\frac{N_0}{A_0 L}\mu_n + \frac{P_0}{A_0 L}\mu_p\right]\frac{A_0}{L} = \frac{e(N_0 \mu_n + P_0 \mu_p)}{L^2} \tag{11.3}$$

where L is the length and A_0 the cross-sectional area of the sample; n_0 and p_0 are the equilibrium concentrations, and N_0 and P_0 are the equilibrium numbers. The change in conductance due to the light is

$$g = \frac{e(\Delta N \mu_n + \Delta P \mu_p)}{L^2} = \frac{eQ(\mu_n \tau_n + \mu_p \tau_p)}{L^2} \tag{11.4}$$

If V is the voltage applied to the photoconductive detector, the change ΔI in current due to the light is

$$\Delta I = V\Delta g = \frac{eV}{L^2}(\mu_n \tau_n + \mu_p \tau_p)Q \tag{11.5}$$

This equation is generally valid; for an electron photoconductor we put $\mu_p = 0$, and for a hole photoconductor we put $\mu_n = 0$.

If the light source is modulated, Q is modulated as well. Q follows the modulation instantaneously because the generation of a hole–electron pair is an extremely fast process, but ΔN and ΔP do not, due to the carrier lifetime.

For an electron photoconductor we have the equation

$$\frac{d\Delta N}{dt} + \frac{\Delta N}{\tau_n} = Q_0 \exp(j\omega t) \tag{11.6}$$

where $Q_0 \exp(j\omega t)$ is the modulated excitation. Trying the solution $\Delta N = \Delta N_0 \exp(j\omega t)$ gives

$$\left(j\omega + \frac{1}{\tau_n}\right)\Delta N_0 = Q_0; \qquad \Delta N_0 = \frac{Q_0 \tau_n}{1 + j\omega \tau_n} \tag{11.7}$$

A similar expression exists for hole–photoconductors. In the general case

Photoconductive Response

we thus have

$$\Delta I = \frac{eV}{L^2}\left[\frac{\mu_n \tau_n}{1+j\omega\tau_n} + \frac{\mu_p \tau_p}{1+j\omega\tau_p}\right]Q_0\exp(j\omega t) \qquad (11.8)$$

For $\omega\tau_n \ll 1$ and $\omega\tau_p \ll 1$ this reduces to (11.5); for $\omega\tau_n \gg 1$ and $\omega\tau_p \gg 1$

$$\Delta I = \frac{eV}{L^2}\left(\frac{\mu_p + \mu_n}{j\omega}\right)Q_0\exp(j\omega t) \qquad (11.8a)$$

so that the response then decreases with increasing frequency. For an electron photoconductor we must again put $\mu_p = 0$ and for a hole photoconductor $\mu_n = 0$.

To illustrate the relationship between photoconductive sensitivity and lifetime we turn to an electron photoconductor. Here

$$\Delta I = \frac{e\mu_n V \tau_n}{L^2} Q \qquad (11.9)$$

However, since

$$\tau_d = \frac{L}{u_d} = \frac{L}{\mu_n |E|} = \frac{L^2}{\mu_n V} \qquad (11.9a)$$

is the *carrier-transit time* (E is the field strength and $u_d = \mu_n |E|$ the drift velocity of the carriers) we may write

$$\Delta I = eQ\frac{\tau_n}{\tau_d} \qquad (11.9b)$$

This has the following simple interpretation. If all carriers were generated at the cathode and collected at the anode, the current would be eQ. The factor $G = (\tau_n/\tau_d)$ is therefore called the *gain* factor; it changes from values much larger than unity for very sensitive, slow photoconductors ($\tau_n \gg \tau_d$) to values much smaller than unity for very insensitive, fast photoconductors ($\tau_n \ll \tau_d$).

The reason why G can be larger than unity is the following. A carrier is not eliminated when it is collected by the anode, but another carrier must enter at the cathode to maintain space charge neutrality. The conduction event that started when a carrier was generated does not stop when the carrier reaches the anode; it only stops when a carrier is removed by recombination or trapping. This does not mean that sensitive photoconductors are necessarily better than insensitive ones. To investigate that problem we must evaluate P_{eq}.

11.1b Noise Response

For an electron photoconductor we had that the fluctuation in the number N of carriers had a spectral intensity

$$S_N(f) = \frac{4g(N_0)}{1/\tau_n^2 + \omega^2} \quad (11.10)$$

whereas according to (11.7)

$$|\Delta N_0| = \frac{Q_0}{(1/\tau_n^2 + \omega^2)^{1/2}} = \frac{\eta(1-R)/eV_{ph}}{(1/\tau_n^2 + \omega^2)^{1/2}} P$$

Equating

$$|\Delta N_0| = [S_N(f)]^{1/2}$$

to find P_{eq}, yields

$$P_{eq} = \frac{eV_{ph}}{\eta(1-R)} [4g(N_0)]^{1/2} \quad (11.11)$$

We see that this is independent of frequency and independent of G. Therefore, sensitive photoconductors do not have a better P_{eq} than insensitive ones; the determining factor is the generation rate $g(N_0)$.

This does not mean that there is no difference between sensitive and insensitive photoconductors. *A more sensitive photoconductor gives a larger signal and therefore the noise of the amplifier handling that signal is less critical. A less sensitive photoconductor is faster, and therefore can handle more information.* Which photoconductor is best depends on the situation.

To investigate $g(N)$ more closely we write

$$g(N_0) = g(n_0) A_0 L \quad (11.12)$$

where $g(n_0)$ is the generation rate per unit volume. It is therefore important to choose photoconductors with a small volume $A_0 L$. Now $A_0 = wd$, where d is the thickness and w, the width of the sample (Fig. 11.1a). Therefore

$$g(N_0) = g(n_0) A d \quad (11.12a)$$

where $A = Lw$ is the receiving area of the photoconductor. Consequently

$$D^* = \frac{A^{1/2}}{P_{eq}} = \frac{\eta(1-R)}{eV_{ph}} \frac{1}{[4g(n_0)d]^{1/2}} \quad (11.13)$$

so that it is important to use thin samples.

Photoconductive Response

Of course d must be so large that practically all light is absorbed, but it should not be made much larger than that. This is especially important if the photoconductivity comes from the excitation of electrons in donor levels. For direct absorption d is of the order of 10^{-5} cm, for excitations from donor levels d may be as small as 10^{-2} cm, depending on the doping.

D^* is inversely proportional to $[g(n_0)]^{1/2}$. It is thus important to reduce $g(n_0)$ by cooling. This is especially important for detectors operating in the far infrared, such as Cu-doped germanium or $Hg_xCd_{1-x}Te$.

11.1c Signal Response of Different Circuits.

One detector circuit is shown in Fig. 11.1b. Here the resistance of the photoconductor is R_d without incident light and $R_d - \Delta R_d$ with incident light; for small amounts of radiation ΔR_d is proportional to the incident power. We thus have from Fig. 11.1b

$$V_R = V_0 \frac{R}{R + R_d}; \qquad V_R + \Delta V_R = V_0 \frac{R}{R + R_d - \Delta R_d}$$

so that

$$\Delta V_R = V_0 \left[\frac{R}{R_d + R - \Delta R_d} - \frac{R}{R_d + R} \right] \cong V_0 \frac{R}{(R_d + R)^2} \Delta R_d \quad (11.14)$$

if $\Delta R_d \ll (R + R_d)$. The maximum signal is obtained if one makes $R = R_d$, in which case

$$\Delta V_R = \frac{1}{4} \frac{V_0}{R_d} \Delta R_d \quad (11.14a)$$

Another possibility is shown in Fig. 11.1c. Here a large d.c. voltage is applied through a large resistance such that $R \gg R_d$. In that case the circuit produces a constant current $I_0 = (V_0/R)$; the output voltage *without* light is $I_0 R_d$ and the output voltage *with* light is $I_0(R_d - \Delta R_d)$. The incident power thus produces an output signal $-I_0 \Delta R_d$ and this signal is increased by making I_0 larger.

In both cases the wanted output signal is proportional to ΔR_d. To distinguish between signal and noise we assume that the signal is described by a resistance variation ΔR_{ds} and the noise, by a resistance fluctuation ΔR_{dn}. Both are multiplied by the same factor, and hence the signal-to-noise power ratio is

$$\frac{S}{N} = \frac{\overline{\Delta R_{ds}^2}}{\overline{\Delta R_{dn}^2}} \quad (11.14b)$$

11.2 EXAMPLES

We now discuss various examples of photoconductors.

11.2a Near-Intrinsic Material; Direct Recombination

In this case, we have $p = (n - N_d)$, where N_d is the donor density

$$g(n) = g_0; \quad r(n) = \rho np = \rho n(n - N_d) \tag{11.15}$$

and the carrier lifetime follows from

$$\tau = \frac{1}{[dr/dn - dg/dn]_{n_0}} = \frac{1}{\rho(2n_0 - N_d)} \tag{11.16}$$

Therefore

$$D^* = \frac{\eta(1-R)}{eV_{ph}} \left(\frac{1}{4g_0 d}\right)^{1/2} \tag{11.16a}$$

Since g_0 varies as $\exp(-eE_g/kT)$, where E_g is the gap width in eV, D^* varies as $\exp(eE_g/2kT)$, so that it is very important to cool the photoconductor.

Direct recombination is forbidden in Ge and Si, but is allowed in low band-gap compounds like InSb and $Cd_xHg_{1-x}Te$. In the first case the lifetime is of the order of 10^{-6} sec.

11.2b Near-Intrinsic Material; Recombination Via Shockley–Read–Hall Centers

In Ge and Si recombination goes via Shockley–Read–Hall recombination centers. What does that do to near intrinsic photoconductors made of these materials? The lifetime now corresponds to the Shockley–Read–Hall lifetime τ_{SR}. The generation rate is in that case

$$g(n_0) = \frac{n_i^2}{(n_0 + n_1)\tau_{p0} + (p_0 + p_1)\tau_{n0}} \tag{11.17}$$

We now take as a simple case intrinsic material in which the trapping level

Examples

lies at the midband level. Then $n = p = n_1 = p_1 = n_i$ and

$$g(n_0) = \frac{n_i}{2(\tau_{p0} + \tau_{n0})} \tag{11.17a}$$

The lifetime τ_{SR} is then $(\tau_{p0} + \tau_{n0})$, $g(n_0)$ varies as n_i, or as $\exp(-eE_g/2kT)$ and hence D varies as $\exp(eE_g/4kT)$. The temperature dependence of D^* is therefore less pronounced than in the previous case, but it is still important to cool the sample.

As an example we consider the case of an intrinsic silicon detector, assuming that it could be made, which has $n_i = 10^{10}/\text{cm}^3$ at room temperature and a lifetime τ of added carriers of 10^{-5} s. We further assume that the detector has a thickness of 0.10 mm and that the received quanta have 1.5 eV energy. We take $\eta = 1.0$ and $R = 0$. What is D^* for this case?

The life time of the added carriers follows from the net Shockley–Read–Hall recombination rate

$$R = \frac{np - n_i^2}{(n+n_1)\tau_{p0} + (p+p_1)\tau_{n0}} = \frac{\Delta n}{\tau} \tag{11.17b}$$

where $n_1 = p_1 = n_i$, $n = p = n_i + \Delta n$, and $\Delta n \ll n_i$. Then

$$R = \frac{2n_i \Delta n}{2n_i(\tau_{p0} + \tau_{n0})}, \quad \text{or} \quad \tau = \tau_{p0} + \tau_{n0} = \tau_{SR} \tag{11.17c}$$

Hence according to (11.17a), $g(n_0) = \frac{1}{2} n_i / \tau = 0.5 \times 10^{15}/\text{cm}^3$ and $D^* = 0.9 \times 10^{12}$ cm$^{1/2}$Hz$^{1/2}$/W.

We see from this example that $g(n_0)$ and τ are not independent but are related, despite the fact that τ does not occur explicity in (11.13).

11.2c Excitation From Donor Levels.

In this case, if N_d is again the donor concentration

$$g(n) = \gamma_0(N_d - n); \quad r(n) = \rho n^2 \tag{11.18}$$

The equilibrium concentration n_0 follows from

$$\gamma_0(N_d - n_0) = \rho n_0^2 \tag{11.18a}$$

and the lifetime τ follows from

$$\tau = \frac{1}{[dr/dn - dg/dn]_{n_0}} = \frac{1}{\gamma_0 + 2\rho n_0} \left(\simeq \frac{1}{\gamma_0} \quad \text{for} \quad 2\rho n_0 \ll \gamma_0 \right) \tag{11.18b}$$

which can be strongly temperature dependent. Finally

$$g(n_0) = \gamma_0(N_d - n_0) \cong \gamma_0 N_d \quad \text{for} \quad n_0 \ll N_d \quad (11.18c)$$

Since γ_0 varies as $\exp(-eE_0/kT)$, where E_0 is the difference in energy between the bottom of the conduction band and the donor level in eV, it is again important to cool the sample. In particular it is important to make $n_0 \ll N_d$; however in that case τ becomes strongly temperature dependent. Since

$$D^* = \frac{\eta(1-R)}{eV_{ph}} \frac{1}{[4\gamma_0 N_d d]^{1/2}} \quad (11.19)$$

we see that the product $N_d d$ must be made so large that practically all of the light is absorbed, but not much larger.

11.2d Excitation From Donor Centers With Counterdoping

Here there are N_d donors/cm^3 and N_a acceptors/cm^3 with $N_a < N_d$. Therefore N_a donors are permanently ionized and at most $N_d - N_a$ donors can absorb energy. Since there are $N_d - N_a - n$ nonionized donors and $N_a + n$ ionized ones,

$$g(n) = \gamma_0(N_d - N_a - n); \quad r(n) = \rho(N_a + n)n \quad (11.20)$$

Therefore the equilibrium concentration follows from

$$\gamma_0(N_d - N_a - n_0) = \rho(N_a + n_0)n_0 \quad (11.20a)$$

$$\tau = \frac{1}{(dr/dn - dg/dn)_{n_0}} = \frac{1}{\gamma_0 + \rho(N_a + 2n_0)} \left(\cong \frac{1}{\rho N_a} \right) \quad (11.20b)$$

if $2n_0 \ll N_a$ and $\gamma_0 \ll \rho N_a$, whereas

$$g_0 = \gamma_0(N_d - N_a - n_0) \cong \gamma_0(N_d - N_a) \quad (11.20c)$$

so that

$$D^* \cong \frac{\eta(1-R)}{eV_{ph}} \frac{1}{[4\gamma_0(N_d - N_a)d]^{1/2}} \quad (11.21)$$

A much faster response can thus be obtained by making N_a large so that $n_0 \ll N_a$ and $\rho N_a \gg \gamma_0$. D^* is then strongly temperature dependent, because γ_0 varies again as $\exp(-eE_0/kT)$, but τ is only temperature dependent in as far as ρ depends on temperature. This can be convenient when one wants detectors with a fast response.

11.3 THERMAL NOISE, FLICKER NOISE, AND AMPLIFIER NOISE

Up to now we assumed that the limiting noise of the photoconductive detector was generation–recombination noise, but in addition there is thermal noise. Can we be so sure that the generation–recombination noise predominates in this case? Can we also be sure that the amplifier noise is negligible?

We answer this question for the circuit of Fig. 11.1c containing an intrinsic semiconductor with band-to-band transitions. Let I_0 be the d.c. current flowing through the photoconductor and let

$$g_0 = \frac{e(\mu_p + \mu_n)N_i}{L^2} = \frac{1}{R_d} \tag{11.22}$$

since $N_0 = P_0 = N_i$, where N_i is the intrinsic number of carriers. The d.c. voltage developed across the detector in the dark is therefore $V_d = I_0 R_d$. The thermal noise gives an open-circuit voltage with a spectrum

$$S_{V_{th}}(f) = 4kTR_d \tag{11.22a}$$

Now let light be shining on the photoconductor, and let this give rise to a change Δg in conductance, then

$$R_d = \frac{1}{g_0}; \quad R_d - \Delta R_d = \frac{1}{g_0 + \Delta g}; \quad \text{then} \quad \Delta R_d = \frac{\Delta g}{g_0^2} = \Delta g R_d^2$$

The change in open-circuit voltage is then

$$\Delta V_d = I_0 \Delta R_d = I_0 R_d^2 \Delta g = V_d R_d \Delta g \tag{11.23}$$

However, since $\Delta N = \Delta P$, in this case

$$\Delta g = \frac{e(\mu_p + \mu_n)}{L^2} \Delta N; \quad \Delta V_d = \left[\frac{e(\mu_p + \mu_n)V_d}{L^2}\right] R_d \Delta N \tag{11.24}$$

and hence the generation–recombination noise has an open-circuit voltage with a spectrum

$$S_{V_{gr}}(f) = \left[\frac{e(\mu_p + \mu_n)V_d}{L^2}\right]^2 R_d^2 S_N(f)$$

$$= \left[\frac{e(\mu_p + \mu_n)V_d}{L^2}\right]^2 R_d^2 2N_i \frac{\tau}{1+\omega^2\tau^2} \tag{11.25}$$

since, according to (5.48)

$$S_N(f) = \overline{4\Delta N^2} \frac{\tau}{1+\omega^2\tau^2}$$

and according to (5.50c), $\overline{\Delta N^2} = \frac{1}{2}N_i$ for intrinsic material. By proper choice of V_d one can always make the generation–recombination noise predominate over the thermal noise, but it becomes more difficult if τ is very small.

To evaluate the effect of flicker noise on the performance of photoconductors, we apply (7.5a). If a voltage V_d is applied to the photoconductor, then the flicker-noise current fluctuation in the current I_0 has a spectrum

$$S_I(f) = \frac{\alpha I_0^2}{N_0 f} \tag{11.26}$$

However,

$$\Delta I = \frac{I_0}{N_0}\Delta N, \quad \text{or} \quad S_I(f) = \frac{I_0^2}{N_0^2}S_N(f) \tag{11.27}$$

Solving for $S_N(f)$ yields

$$[S_N(f)]_{\text{flicker}} = \frac{\alpha N_0}{f} \tag{11.28}$$

This must now be compared with the value of $S_N(f)$ for generation–recombination noise. We had for nearly intrinsic material

$$[S_N(f)]_{gr} = \frac{4P_0 N_0}{N_0+P_0}\frac{\tau}{1+\omega^2\tau^2} \tag{11.29}$$

where τ is the lifetime of the recombination process and $P_0 \simeq N_0$. We thus see that flicker noise can become quite important in fast, near-intrinsic photoconductors (τ small). Since both $S_n(f)$ values are multiplied by the same factor to obtain the $S_V(f)$ values, the relative effect of flicker noise is adequately described by (11.28) and (11.29).

If flicker noise is present, one can always define a frequency f_c, the so-called *upper cut-off frequency*, for which $[S_N(f)]_{\text{flicker}} = [S_N(f)]_{gr}$; this frequency is independent of bias. For $f < f_c$ flicker noise predominates and for $f > f_c$ generation–recombination noise predominates.

It should be taken into account that for practical detectors D^* depends on thermal noise and flicker noise. At a given bias D^* increases with increasing frequency until flicker noise becomes smaller than the combined

Noise-Reduction Methods

effect of thermal noise and generation–recombination noise. If now the bias is increased, the relative effect of thermal noise becomes smaller and D^* increases further until generation–recombination noise predominates over thermal noise, if that point can be reached without penalty.

We now investigate the effect of amplifier noise. To that end we introduce the noise figure F_0 of the photoconductor by equating

$$S_{V_{gr}}(f) = F_0 S_{V_{th}}(f) \text{ or } F_0 = \frac{e}{2kT} \frac{(\mu_p + \mu_n)V_d^2}{L^2} \frac{\tau}{1+\omega^2\tau^2} \quad (11.30)$$

as is found by substituting for R_d.

Let for the source resistance R_d the noise figure of the amplifier be F, then the noise of detector plus amplifier for a bandwidth Δf has a mean square value

$$[(F_0 + F)\cdot 4kTR_d\Delta f]^{1/2} = [F_1 \cdot 4kTR_d\Delta f]^{1/2} \quad (11.31)$$

so that the overall noise figure F_1 of detector plus amplifier is

$$F_1 = F_0 + F \quad (11.32)$$

It is thus important to have $F < F_0$; this can be achieved by making V_d sufficiently large.

One should be careful, however, that the voltage V_d is so chosen that the biased device still has a linear characteristic. The characteristic can become nonlinear because of carrier injection at the contacts. This can give rise to a large amount of low-frequency noise; hence such a situation must be avoided.

11.4 NOISE-REDUCTION METHODS

We see in this section how noise in photoconductive detector systems can be reduced by phase-sensitive detection methods.

We saw in Section 8.2b that a phase-sensitive detector had a very small effective bandwidth B_{eff}. In photoconductive detector systems this can be utilized by chopping the light beam at a frequency f and detecting the modulated output with a phase-sensitive detector synchronized with the chopper. If the detector has an area A and a detectivity D^*, then the effective noise power P_{eff} of the system is

$$P_{\text{eff}} = \frac{A^{1/2}}{D^*} B_{\text{eff}} \quad (11.33)$$

so that a lower P_{eff} can be obtained by making B_{eff} small.

The chopper frequency f and the device bias should be so chosen that D^* has its maximum value. This means that $f > f_c$.

Further improvement is possible by feeding the photoconductor from a d.c. source V_d and an a.c. source $V_d \cos \omega_0 t$ in series, where the frequency f_0 is so chosen that $f_0 \ll f$. In that case the signal, the generation–recombination noise, and the flicker noise are modulated at the frequency f_0, whereas thermal noise is not modulated. If the signal is detected by a phase-sensitive detector synchronized with the modulating voltage of frequency f_0, the thermal noise disappears and only flicker and generation–recombination noise remain. However, flicker noise can be reduced to a negligible amount by proper choice of the chopper frequency f, so that only the generation–recombination noise remains as the essential noise source.

To put this method in effect, the output signal of the photoconductor should be passed through a high-pass filter, amplified by a tuned amplifier tuned at the frequency f, detected by a linear detector, and the detected output should be fed into a phase-sensitive detector driven by the modulating voltage of frequency f_0 (Fig. 11.2).

It should be understood, of course, that these methods have limited applicability. Quite often one is interested in receiving light signals with wideband modulation. In that case one must use wideband amplifiers to process the detected signals.

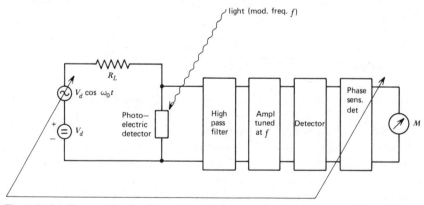

Figure 11.2. Photoconductive detector arrangement in which a d.c. bias and an a.c. bias of frequency ω_0 are used. By using a phase-sensitive detector driven by the a.c. bias signal of frequency ω_0 the photodiode response is detected and the thermal noise of the photoconductor is eliminated.

11.5 PRACTICAL EXAMPLES OF PHOTOCONDUCTORS

In practice every semiconducting element can be used as a photoconductor. In particular Ge and Si can be used as photoconductors in the near infrared. Also various III-V compounds such as InSb can be used.

The compound InSb has a photoconductive response below 5.5 μm wavelength at a temperature of 77 ° K. It has a D^* between 10^{10} and 10^{11} cm $Hz^{1/2}/W$. At higher temperatures the D^* is of course less, since the intrinsic conductivity of the material increases with increasing temperature.

Materials such as $Pb_{1-x}Sn_xTe$ and $Pb_{1-y}Sn_ySe$ have a gap width that depends on x and y, respectively. By proper choice of x or y one can make photoconductive detectors that can operate in the 8–15 μm wavelength range, where the atmosphere is transparent. Peak detectivities D of a few times 10^8 cm $Hz^{1/2}/W$ are feasible, with considerable improvement in D^* by going to 4.2° K.

Similar results have been obtained for $Hg_{1-x}Cd_xTe$. Here gap widths between 0 and 0.40 eV are obtained for $0.17 < x < 0.40$ at 0° K. The photoconductive devices made of this material have a certain amount of flicker noise at low frequencies but have otherwise excellent characteristics.

Doped germanium and silicon materials are also useful. Of special interest is copper-doped germanium, which has an energy level 0.04 eV above the valence band. The material has excellent photoconductive characteristics around 10 μm.

The preceding examples are only a few of the many possibilities available.

REFERENCE

R. K. Willardson and A. C. Beer (eds.), *Semiconductors and Semimetals, Vol. 5, Infrared Detectors*, Academic, New York, 1970: especially Chapter 2, P. W. Kruse, Indium antimonide photoconductive and photoelectromagnetic detectors; Chapter 4, I. Melngailis and T. C. Harman, Single-crystal lead–tin chalcogenides; Chapter 5, D. Long and J. L. Schmit, Mercury–cadmium telluride and closely related alloys; Chapter 8, R. J. Keyes and T. M. Quist, Low-level coherent and incoherent detection in the infrared.

12

PYROELECTRIC DETECTORS AND CAPACITIVE BOLOMETERS

A pyroelectric detector is a detector based on the *pyroelectric effect* that is, on a change in polarization in a poled ferroelectric capacitor produced by heating due to incident radiation. The reason is that in a poled ferroelectric material the saturation polarization P_s depends on temperature. A temperature variation thus causes a change in polarization which in turn results in a change of charge on the capacitor and thus produces a change in voltage that can be processed by standard techniques (Section 12.1).

The small-signal capacitance of a ferroelectric capacitor operating in the paraelectric mode also depends strongly on temperature. A change in temperature, such as can be produced by incident radiation, thus gives rise to a change in capacitance; if d.c. bias is applied, a change in output voltage results that can be processed by standard techniques. This is called the d.c. capacitive bolometer. In the a.c. capacitive bolometer the capacitance change is detected by an a.c. technique (Section 12.2). Section 12.3 discusses dielectric losses as polarization fluctuations.

12.1 THE PYROELECTRIC DETECTOR

A pyroelectric detector receives modulated radiation $P_1 \exp(j\omega t)$ that can be obtained, for instance, by chopping the radiation (Fig. 12.1). The detector has a thickness w and an area A. In the process a voltage $v_{d0}\exp(j\omega t)$ is developed across the terminals. The front face of the capacitor is often painted black to absorb the radiation.

We shall now use the following symbols to describe the significant parameters. The small-signal device capacitance is $C = (\varepsilon\varepsilon_0 A/w)$, where ε_0 is the dielectric constant of free space and ε the small-signal relative

The Pyroelectric Detector

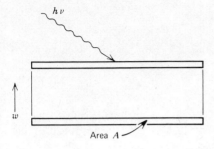

Figure 12.1. Pyroelectric detector of area A and thickness w, receiving modulated light of frequency ω and producing an a.c. voltage $v_{d0}\exp(j\omega t)$ across the terminals.

dielectric constant of the material. The device has a heat capacitance $C_H = cdAw$, c being the specific heat per gram and d the density of the material. The front face has a heat loss conductance $g_H = \eta.4\sigma T^3 A$ by radiation, and the back face has a heat-loss conductance $g'_H = \eta'.4\sigma T^3 A$, where η and η' are the emissivities of front and back face, respectively, and σ is the Stefan-Boltzmann constant.

12.1a Technical Sensitivity and NEP P_{eq}

We start with calculating the technical sensitivity. The heat response of the detector due to incident radiation $P_1 \exp(j\omega t)$ is given by

$$C_H \frac{d\Delta T}{dt} + (g_H + g'_H)\Delta T = P_1 \exp(j\omega t) \qquad (12.1)$$

Putting $\Delta T = \Delta T_0 \exp(j\omega t)$ and solving for ΔT_0 yields

$$\Delta T_0 = \frac{\eta P_1}{j\omega C_H + g_H + g'_H} \qquad (12.2)$$

The thermal time constant $\tau_H = C_H/(g_H + g'_H)$ is generally of the order of 1 s and ω is usually of the order of 10–100 s^{-1} so that $\omega^2 \tau_H^2 \gg 1$. Therefore $\omega^2 C_H^2 \gg (g_H + g'_H)^2$, or

$$\Delta T_0 \cong \frac{\eta P_1}{j\omega C_H} = \frac{P_1}{j\omega cdAw} \qquad (12.2a)$$

Now the polarization P_s of the polarized ferroelectric material decreases with increasing temperature. The charge on the capacitor is $Q = AP_s$ and hence the short-circuited current is

$$I_d(t) = \frac{dQ}{dt} = A\frac{dP_s}{dt} = A\frac{dP_s}{dT} \cdot \frac{dT}{dt} = -Apj\omega \Delta T_0 \exp(j\omega t) = -I_{d0}\exp(j\omega t) \tag{12.3}$$

where $p = (-dP_s/dT)$ is called the pyroelectric coefficient. Since the a.c. current is proportional to $d\Delta T/dt$, rather than to ΔT, the response of the detector is very fast. The current amplitude

$$I_{d0} = Apj\omega \Delta T_0 = \frac{p\eta P_1}{cdw} \tag{12.3a}$$

is therefore independent of frequency for $\omega \tau_H \gg 1$.

The voltage developed across the capacitor C has an amplitude

$$V_{d0} = \frac{I_{d0}}{\omega C} = \frac{p\eta}{\omega c d \varepsilon \varepsilon_0 A} P_1 \tag{12.4}$$

The *voltage response* thus decreases with increasing frequency. Equation (12.4) gives the technical sensitivity V_{d0}/P_1; it is expressed in V/W.

We now calculate P_{eq}. The limiting noise of the device is generally the noise of dielectric losses. Its spectral intensity is

$$S_i(f) = 4kT\omega C \tan\delta = 4kT\omega \frac{\varepsilon \varepsilon_0 A}{w} \tan\delta \tag{12.5}$$

where $g = \omega C \tan\delta$ is the loss conductivity of the capacitor and $\tan\delta$ is its loss factor. The noise equivalent power, P_{eq}, defined by

$$I_{d0} = [S_i(f)]^{1/2} \tag{12.6}$$

yields

$$P_{eq} = \frac{[S_i(f)]^{1/2} cdw}{p\eta} = \frac{cd}{p\eta}(4kT\varepsilon\varepsilon_0 \tan\delta)^{1/2}(\omega wA)^{1/2} \tag{12.7}$$

Consequently D^* may be expressed as

$$D^* = \frac{A^{1/2}}{P_{eq}} = \frac{p\eta}{cd} \frac{1}{(4kT\varepsilon\varepsilon_0 \tan\delta)^{1/2}(\omega w)^{1/2}} \tag{12.7a}$$

Since D^* varies as $w^{-1/2}$, it pays to go to thinner samples; it is also important to keep ω relatively small.

The Pyroelectric Detector

The figure of merit of the detector is therefore

$$\frac{p}{(\varepsilon \tan \delta)^{1/2}} \tag{12.7b}$$

Can this be improved by going to a different temperature? Usually not, for $p/\varepsilon^{1/2}$ is nearly independent of temperature, and $\tan \delta$ is not strongly temperature dependent. Can one improve D^* by going to a different material? Unfortunately not, for $p/\varepsilon^{1/2}$ turns out to be nearly independent of the material.

To find the order of magnitude for P_{eq} and D^* we take triglycine sulfate at 25° C. Here $\varepsilon = 40$, $\tan \delta = 0.01$, $c = 0.97$ joule deg^{-1} g^{-1}, $d = 1.69$ g cm^{-3}, and $p = 2.0 \times 10^{-8}$ coulomb cm^{-2} deg^{-1}. If we put, furthermore, $\eta = 1$, $\omega = 100$ s^{-1}, $A = 1$ mm^2, $w = 10^{-2}$ cm, and $\varepsilon_0 = 8.85 \times 10^{-14}$ F cm^{-1}, then

$$P_{eq} = 2.0 \times 10^{-10} \text{ W/Hz}^{1/2}$$

so that

$$D^* = \frac{A^{1/2}}{P_{eq}} = 5 \times 10^8 \text{ cm Hz}^{1/2}/\text{W}$$

This is reasonably good, which explains why pyroelectric detectors have attracted so much interest, but is far below the limit set by spontaneous temperature fluctuation noise. This makes it highly unlikely that ferroelectric materials will ever be found that will come much closer to the temperature-fluctuation noise limit.

12.1b The Factor $p/\varepsilon^{1/2}$.

To demonstrate that $p/\varepsilon^{1/2}$ is independent of T we turn to Devonshire's theory of ferroelectrics. According to this theory the field E is a nonlinear function of the polarization P (see Appendix)

$$E = \beta(T - T_c)P + B_0 P^3 + C_0 P^5 + \cdots \tag{12.8}$$

where T_c is the Curie temperature of the material and β, B_0, and C_0 are constants. Therefore, for $T < T_c$ there is spontaneous polarization, namely, the equation $E = 0$ has a nonzero solution for P. Neglecting the P^5 term we thus have

$$\beta(T - T_c)P + B_0 P^3 = 0; \quad P = P_s = \left[\frac{\beta(T_c - T)}{B_0}\right]^{1/2} \tag{12.9}$$

so that P_s decreases with increasing T. Hence

$$p = -\frac{\partial P_s}{\partial T} = \frac{1}{2}\left(\frac{\beta}{B_0}\right)^{1/2}\frac{1}{(T_c - T)^{1/2}} \qquad (12.9a)$$

The differential susceptibility is

$$\varepsilon - 1 = \frac{1}{\varepsilon_0 \partial E/\partial P} = \frac{1/\varepsilon_0}{\beta(T - T_c) + 3B_0 P_s^2} = \frac{1}{2\beta\varepsilon_0(T_c - T)} \qquad (12.9b)$$

so that

$$\frac{p}{(\varepsilon - 1)^{1/2}} \cong \frac{p}{\varepsilon^{1/2}} = \frac{\frac{1}{2}(\beta/B_0)^{1/2}}{1/(2\beta\varepsilon_0)^{1/2}} = \frac{\varepsilon_0^{1/2}\beta}{(2B_0)^{1/2}} \qquad (12.10)$$

which is independent of the temperature. Since $\varepsilon \gg 1$ for these materials, $\varepsilon - 1 \cong \varepsilon$. If the $C_0 P^5$ term is taken into account, one finds a moderate dependence of $p/(\varepsilon - 1)^{1/2}$ upon $T_c - T$. Since (12.8) converges only for relatively small P, (12.10) is only accurate if $T_c - T$ is not large.

To prove that $p/(\varepsilon - 1)^{1/2}$ is practically independent of the material, we evaluate β and B with the help of the molecular field model. We assume that the elementary cells in the ferroelectric material can be polarized parallel and antiparallel to the local electric field and have a dipole moment μ. The local field E_l acting on this elementary dipole μ is

$$E_l = E + \frac{\lambda P}{\varepsilon_0} \qquad (12.11)$$

where λ is the Lorentz factor of the material (see Appendix).

It is easily shown that the net polarization P for N elementary cells per unit volume is

$$P = N\mu \tanh x; \qquad x = \frac{\mu E_l}{kT} \qquad (12.12)$$

The proof is as follows. The probability of a parallel orientation is $A \exp(\mu E_l/kT)$ and the probability for an antiparallel orientation is $A \exp(-\mu E_l/kT)$. Now A must be so chosen that the sum of the two probabilities is unity, or

$$A = \left[\exp\left(\frac{\mu E_l}{kT}\right) + \exp\left(-\frac{\mu E_l}{kT}\right)\right]^{-1}$$

whereas the net polarization is $N\mu[A \exp(\mu E_l/kT) - A \exp(-\mu E_l/kT)]$
Putting it all together yields (12.12).

We now invert (12.12) as follows

$$x = \frac{\mu E_l}{kT} = \frac{P}{\mu N} + \frac{1}{3}\left(\frac{P}{\mu N}\right)^3 + \cdots \qquad (12.13)$$

or

$$E + \frac{\lambda P}{\varepsilon_0} = \frac{kT}{\mu^2 N} P + \frac{kT}{3\mu^4 N^3} P^3 + \cdots \qquad (12.13a)$$

Writing E in the form (12.8) yields

$$\beta = \frac{k}{\mu^2 N}; \quad B_0 = \frac{kT_c}{3\mu^4 N^3}; \quad T_c = \frac{\lambda}{\varepsilon_0} \frac{\mu^2 N}{k} \qquad (12.14)$$

where we have replaced T by T_c in B_0; this is done because the inversion (12.13) is only valid for T close to T_c. Consequently (12.8) has now been proved, and

$$\frac{p}{(\varepsilon-1)^{1/2}} = \left(\frac{3}{2}\varepsilon_0 \frac{Nk}{T_c}\right)^{1/2} \qquad (12.15)$$

We must now find a suitable expression for the density of elementary cells N. It is easily seen that

$$N = \frac{A_0 \rho}{nW} \qquad (12.16)$$

where A_0 is Avogadro's number, ρ is the density, W is the molecular weight of each molecule, and n is the number of molecules per unit cell. This yields

$$\frac{p}{(\varepsilon-1)^{1/2}} = 10.5 \times 10^{-9} \left\{\frac{\rho}{n \cdot \frac{W}{100} \cdot \frac{T_c}{100}}\right\}^{1/2} \text{coulomb cm}^{-2\circ} \text{K}^{-1} \qquad (12.17)$$

This usually lies within $(3-8) \times 10^{-9}$ coulomb cm$^{-2\circ}$ K^{-1}, whereas the experimental values lie around 3×10^{-9}. Considering the rather crude model, this is good agreement. We can thus understand why $p/\varepsilon^{1/2}$ is practically independent of the material.

12.1c Noise in Pyroelectric Arrays

If one has an array of pyroelectric detectors, and one scans them at the rate f_1, say 10–30 s^{-1}, then one must take into account the noise generated in a band of frequencies. The lowest frequency is of course f_1, since the low-frequency noise does not change enough between two scans to be recorded. The highest frequency $f_2 \simeq 1/\tau_s$, where τ_s is the scanning time of a single element, since the high-frequency noise is averaged over the scanning time. We thus have for frequency-independent $\tan\delta$

$$S_v(f) = 4kT \int_{f_1}^{f_2} \frac{\tan\delta}{\omega C} df = \frac{2}{\pi} \frac{kT}{C} \tan\delta \ln\left(\frac{f_2}{f_1}\right) \qquad (12.18)$$

Since $S_v(f)$ depends only logarithmically on f_2 and f_1, any error in our estimate of f_2/f_1 has little effect.

We thus see that the noise observed can be much less than kT/C, the value expected from the equipartition law, especially if $\tan\delta$ is relatively small. The reason is twofold:

1. The equipartition law does not hold for the voltage developed across lossy dielectrics*.

2. We do not integrate over the full frequency range $0 \leqslant f < \infty$ but only over the range $f_1 \leqslant f \leqslant f_2$.

12.2 THE CAPACITIVE BOLOMETER

The capacitance variation due to heating of a capacitor can be detected with the help of a d.c. bias by an applied voltage V_0 or by means of an a.c. bias voltage $V_0 \cos\omega_0 t$, where ω_0 is much larger than the modulation frequency ω of the incident radiation. We call this, respectively, the d.c. biased bolometer and the a.c. biased bolometer. We shall see that the latter may have some advantages as far as noise is concerned.

12.2a The D. C. Biased Bolometer

We investigate the operation of the device for $T > T_c$. The material is then paraelectric and $C = \text{const}/(T - T_c)$. Modulated radiation of frequency ω produces a variation in capacitance; this variation is detected by applying a d.c. voltage V_0 in series with the capacitor and connecting the circuit to a large load resistance R_L so that $\omega C_0 R_L \gg 1$, where C_0 is the equilibrium value of C (Fig. 12.2).

*A. van der Ziel, *J. Appl. Phys.*, **44**, 1400 (1973); **44**, 1402 (1973).

The Capacitive Bolometer

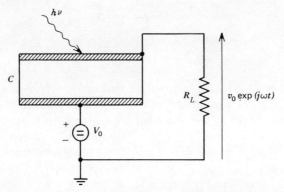

Figure 12.2. Capacitive bolometer receiving modulated light. When driven by a d.c. supply voltage V_0 a signal voltage $v_0 \exp(j\omega t)$ is developed across a load resistor R_L.

If the temperature variation is $\Delta T_0 \exp(j\omega t)$ the capacitance may be written

$$C = C_0 + \frac{\partial C_0}{\partial T} \Delta T_0 \exp(j\omega t) \quad (12.19)$$

However, the charge Q of the capacitor is CV_0 if the capacitor is linear, so that the short-circuit current is

$$I = \frac{dQ}{dt} = V_0 \frac{dC}{dt} = j\omega V_0 \frac{\partial C_0}{\partial T} \Delta T_0 \exp(j\omega t) \quad (12.20)$$

Hence the output voltage $v_0 \exp(j\omega t)$ has an amplitude

$$v_0 = \frac{j\omega V_0 (\partial C_0/\partial T) \Delta T_0}{j\omega C_0} = V_0 \frac{1}{C_0} \frac{\partial C_0}{\partial T} \Delta T_0 = -\frac{\eta P_1 V_0/w}{j\omega c dA(T-T_c)} \quad (12.21)$$

so that the output voltage is proportional to the d.c. field V_0/w. We have here substituted (12.2a) for ΔT_0.

Assuming again the noise of the device to be thermal noise of the dielectric losses, we have

$$S_v(f) = \frac{4kT \tan \delta}{\omega C_0} = \frac{4kTw \tan \delta}{\omega \varepsilon \varepsilon_0 A} \quad (12.22)$$

Defining the NEP, P_{eq}, by

$$(v_0)_{r.m.s.} = [S_v(f)]^{1/2}$$

yields

$$P_{eq} = \frac{cd(T-T_c)}{\eta V_0/w} \left(\frac{4kT\tan\delta}{\varepsilon\varepsilon_0}\right)^{1/2} (\omega Aw)^{1/2} \qquad (12.23)$$

The dependence on ω, A, and w is the same as for the pyroelectric detector. Because of the dependence on ε, it is important to use materials with a large relative dielectric constant ε. Moreover, since P_{eq} varies as $(T-T_c)^{3/2}$ ($1/\varepsilon$ varies as $T-T_c$), it is important to keep $T-T_c$ small. Finally P_{eq} varies inversely with V_0/w, so that it is important to go to high fields. One cannot go too far in this direction, however, since saturation effects set in at high fields (Section 12.2b).

As an example, we consider here a capacitive bolometer made with triglycine sulfate, which is not a good material, since ε is relatively small. Substituting $cd = 1.6$ joule/cm^3, $T = 320°$ K, $\eta = 1$, $\omega = 100/s$, $V_0/w = 2.5$ kV/cm, $T-T_c = 1.0°$ C, $\tan\delta = 0.01$, $\varepsilon = 800$, $A = 1$ mm^2, $\varepsilon_0 = 8.85 \times 10^{-14}$ F/cm, $w = 10^{-2}$ cm, $P_{eq} = 1.0 \times 10^{-10}$ W/Hz$^{1/2}$, which is comparable with the findings obtained for the triglycine sulfate pyroelectric detector. Materials with a much larger value of ε should be considerably better than this.

The experiments agree with the theoretical predictions as far as order of magnitude is concerned, but $\tan\delta$ can be somewhat larger near the Curie temperature and increases with increasing applied field strength.* Finally, the noise of the device under d.c. bias is somewhat larger than thermal noise, which is not so surprising, since the device is operating under nonequilibrium conditions.

12.2b Saturation Effects

We now investigate how saturation effects can be taken into account. We cannot use the small-signal a.c. capacitance C directly, and the charge Q is no longer equal to CV_0. Rather, one must introduce an induced polarization $P(V_0, T)$ and put $Q = AP(V_0, T)$. The current is therefore

$$I(t) = \frac{dQ}{dt} = j\omega A \frac{\partial P}{\partial T} \Delta T_0 \exp(j\omega t) = I_{d0}\exp(j\omega t) \qquad (12.24)$$

and the output voltage has an amplitude

$$v_0 = \frac{I_{d0}}{j\omega C_0} = \frac{A}{C_0}\frac{\partial P}{\partial T}\Delta T_0 \qquad (12.25)$$

*V. P. Singh, Ph. D. Thesis, University of Minnesota, 1973.

The Capacitive Bolometer

Since the small signal capacitance C_0 is given as

$$C_0 = \frac{\partial Q}{\partial V_0} = A \frac{\partial P}{\partial V_0} \tag{12.26}$$

we may write

$$v_0 = \frac{\partial P/\partial T}{\partial P/\partial V_0} \Delta T_0 \tag{12.25a}$$

This expression for v_0 corresponds exactly to the pyroelectric case. The capacitive bolometer effect can thus be interpreted as an *induced pyroelectric effect*.

To show that (12.25) can still be interpreted as a *capacitive* bolometer effect, we introduce the d.c. capacitance $C_d = (Q/V_0)$, or $Q = AP = C_d V_0$ or $P = (C_d V_0/A)$. Hence

$$v_0 = V_0 \frac{1}{C_0} \frac{\partial C_d}{\partial T} \Delta T_0 \tag{12.27}$$

The only correction needed for (12.21) is that the d.c. capacitance C_d rather than the a.c. small signal capacitance C_0 should be considered as the temperature-sensitive element. For the linear case (12.27) reduces to (12.21) since $C_d = C_0$. Saturation means that $(V_0/C_0)(\partial C_d/\partial T)$ saturates for large V_0.

12.2c The A. C. Biased Capacitive Bolometer

In the a.c. biased bolometer one applies an a.c. signal $V_0 \cos\omega_0 t$ to the bolometer, where ω_0 is much larger than the frequency ω of the incident radiation, and finds the response. It now turns out that P_{eq} is lower by a factor $(\omega/\omega_0)^{1/2}$.

To show this, we turn to the circuit of Fig. 12.3, where the temperature-dependent capacitance C is tuned to the carrier frequency ω_0. To obtain the sideband frequencies, we let C vary as $[C_0 + C_1 \exp(\pm j\omega t)]$. Hence if $\omega_0^2 L C_0 = 1$,

$$\begin{aligned} i_0 &= \frac{v_i}{j\omega_0 L + r_L + r_c + 1/\{j\omega_0[C_0 + C_1 \exp(\pm j\omega t)]\}} \\ &\cong \frac{v_i j\omega_0 C_0}{1 - \omega_0^2 L[C_0 + C_1 \exp(\pm j\omega t)] + (r_L + r_c)j\omega_0 C_0} \\ &= \frac{v_i j\omega_0 C_0}{j/Q_0 - (C_1/C_0)\exp(\pm j\omega t)} = -\frac{jQ_0 j\omega_0 C_0 v_i}{1 + j(C_1/C_0)Q_0 \exp(\pm j\omega t)} \end{aligned} \tag{12.28}$$

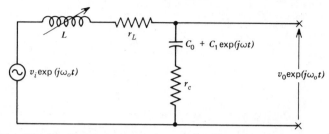

Figure 12.3. Capacitive bolometer circuit driven by an a.c. signal $v_i \exp(j\omega_0 t)$ and developing a modulated signal of frequency $\omega_0 = (\omega_i \pm \omega)$, where ω is the modulation frequency of the incoming radiation.

where $1/Q_0 = \omega_0 C_0 (r_L + r_c)$ and Q_0 is the Q factor of the tuned circuit. If we now choose the amplitude v_i such that $Q_0 v_i = V_0$, we may write

$$v_0 = \frac{i_0}{j\omega_0 C_0} = -jV_0 \left[1 - jQ_0 \frac{C_1}{C_0} \exp(\pm j\omega t) \right] \quad (12.29)$$

so that the signal $v_0 \exp(j\omega_0 t)$ contains a carrier of frequency ω_0 and two sidebands of frequencies $\omega_0 + \omega$ and $\omega_0 - \omega$, respectively.

After detection by a linear detector we have, if v_0', ΔT_0, and P_1 are r.m.s. values, since $C_1 = (\partial C_0/\partial T)\Delta T_0$ and $(\partial C_0/\partial T)/C_0 = 1/(T - T_c)$, and having substituted for ΔT_0

$$v_0' = V_0 Q_0 \frac{\Delta T_0}{T - T_c} = \frac{Q_0 \eta V_0/w}{cdA(T - T_c)} P_1 \quad (12.30)$$

where cd is the specific heat per cm^3, w the device thickness, and A the device area.

The noise per unit bandwidth is

$$\overline{(v_n^2)}^{1/2} = Q_0 [4kT(r_L + r_c)]^{1/2} = Q_0 \left(4kT \frac{\tan \delta_0}{\omega_0 C_0} \right)^{1/2} \left(1 + \frac{r_L}{r_c} \right)^{1/2} \quad (12.31)$$

where $\tan \delta_0$ is the loss factor at $\omega = \omega_0$. Therefore, we have for $(P_{eq})_{a.c.}$

$$(P_{eq})_{a.c.} = \frac{cd(T - T_c)}{V_0/w} \left(\frac{4kT \tan \delta_0}{\varepsilon \varepsilon_0} \right)^{1/2} (\omega A w)^{1/2} \left(\frac{\omega}{\omega_0} \right)^{1/2} \left(1 + \frac{r_L}{r_c} \right)^{1/2} \quad (12.32)$$

Comparing this with the corresponding expression $(P_{eq})_{d.c.}$ for the d.c.

The Capacitive Bolometer

biased bolometer at the same value of V_0, since $\tan\delta_0 \cong \tan\delta$

$$(P_{eq})_{a.c.} = (P_{eq})_{d.c.}\left(\frac{\omega}{\omega_0}\right)^{1/2}\left(1+\frac{r_L}{r_c}\right)^{1/2} \qquad (12.32a)$$

If we use the same example as in the previous section and put $\omega/\omega_0 = 1/1000$ and assume $r_L \ll r_c$, we have

$$(P_{eq})_{a.c.} = \frac{1.2 \times 10^{-10}}{31.6} \text{W}/\text{Hz}^{1/2} = 4 \times 10^{-12} \text{W}/\text{Hz}^{1/2}$$

which is comparable to the limit set by temperature-fluctuation noise. Strictly speaking, we should write here

$$(P_{eq})_{a.c.} = \left[(P_{eq})^2_{\text{diel}} + (P_{eq})^2_{\text{temp}}\right]^{1/2} \qquad (12.33)$$

where $(P_{eq})_{\text{diel}}$ and $(P_{eq})_{\text{temp}}$ are the NEPs due to dielectric losses and due to temperature-fluctuation noise, respectively.

The circuit of Fig. 12.3 has the good feature of amplifying both the signal and the noise by the factor Q_0, but it has the undesirable feature that the carrier voltage is very large. One can improve this by using the balanced circuit of Fig. 12.4 and detect the modulated signal without carrier with the help of a phase-sensitive detector.

Since half the signal is lost by the balanced circuit and the r.m.s. noise is $2^{-1/2}$ times the value found in the previous case, P_{eq} is a factor $2^{1/2}$ larger than in the previous case. In view of the great advantage of the balanced

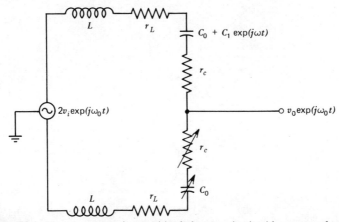

Figure 12.4. An a.c. symmetric capacitive bolometer circuit with suppressed carrier.

circuit this increase in P_{eq} should be tolerable. To improve the balance, one should use a capacitive bolometer in each branch of the circuit, and only irradiate one of them with infrared light. This also helps in overcoming the effect of the instantaneous voltage on the device capacitance, a problem that has not yet been solved but that can be reduced by keeping V_0 relatively small.

The discussion given here is probably somewhat too optimistic. In the first place one would expect the noise to be somewhat larger than thermal noise of the dielectric losses, since we have a nonequilibrium situation. In the second place the circuit is a mixer that might possibly take noise at the frequency ω and turn it into excess noise at the sideband frequencies $\omega_0 \pm \omega$. This complicated mixing problem has not yet been solved. Last, but probably not least, there is a noise effect due to the motion of domain walls (Barkhausen noise); it can be kept small by keeping V_0 relatively small.

12.3 THE NATURE OF THE DEVICE NOISE

We shall now discuss the nature of the device noise. To that end we turn to Fig. 12.5, which shows a d.c. biased ferroelectric capacitor. If Q is the charge on the capacitor, A the electrode area, and w the thickness of the capacitor, we have for the dielectric displacement D:

$$D = \frac{Q}{A} = \varepsilon_0 E + P \qquad (12.34)$$

Since E is kept constant and the fluctuation in Q produces the device

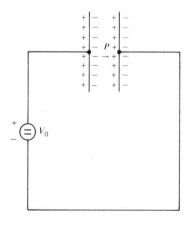

Figure 12.5. A d.c. biased ferroelectric capacitor showing a polarization P.

noise, we see that the only source of noise is the fluctuation in P. We thus investigate this *polarization noise* more closely.

Since $Q = A(P + \varepsilon_0 E)$, we have

$$I(t) = \frac{d}{dt}(AP) \tag{12.35}$$

Making a Fourier analysis, we obtain

$$i_n = j\omega A P_n, \quad \text{or} \quad S_I(f) = \omega^2 A^2 S_p(f) \tag{12.35a}$$

We next consider the problem for zero applied field. Then

$$S_I(f) = 4kT(\omega C_0 \tan\delta), \quad \text{or} \quad S_p(f) = \frac{4kT}{\omega A^2} C_0 \tan\delta \tag{12.36}$$

Now we put $\varepsilon = (\varepsilon' + j\varepsilon'')$, then $C_0 = \varepsilon' \varepsilon_0 A/w$ and $\tan\delta = (\varepsilon''/\varepsilon')$, where ε' and ε'' are the real and imaginary parts of the complex dielectric constant ε, respectively; here ε'' will depend on frequency. Then

$$S_p(f) = \frac{4kT\varepsilon_0 \varepsilon''(\omega)}{\omega V} \tag{12.36a}$$

where $V = Aw$ is the volume of the dielectric. Hence the total polarization fluctuation is

$$\overline{\Delta P^2} = \int_0^\infty S_p(f) df = \frac{4kT\varepsilon_0}{V} \int_0^\infty \frac{\varepsilon''(\omega) df}{\omega} = \frac{kT\varepsilon_0(\varepsilon_s - 1)}{V} \tag{12.37}$$

where ε_s is the static dielectric constant, for according to the Kramers–Kronig relation for dielectrics

$$\varepsilon_s - 1 = \frac{2}{\pi} \int_0^\infty \frac{\varepsilon''(\omega) d\omega}{\omega} \tag{12.37a}$$

Equation (12.36a) is correct at equilibrium, that is, at zero field.

In nonequilibrium situations ε'' becomes dependent on the electric field, but in addition extra noise may be generated. This problem has not yet been solved theoretically.

REFERENCE

E. H. Putley, The pyroelectric detector, hn R. K. Willardson and A. C. Beer (eds.), *Semiconductors and Semimetals*, Vol. 5, Academic, New York, 1970, Chapter 6.

13

NOISE IN TELEVISION PICKUP TUBES

In this chapter we discuss various television pickup tubes, such as the image orthicon (Section 13.1), the vidicon (Section 13.2), and the secondary electron conduction vidicon (Section 13.3). Section 13.4 discusses scanned-image sensors such as multiplexed photodiode arrays and charge-coupled image sensors. Section 13.5 discusses the pyroelectric vidicon.

13.1 THE IMAGE ORTHICON

The image orthicon is shown in Fig. 13.1. It consists of a photocathode, a glass target with a collector electrode in front of it, an electron beam scanning the target and reestablishing its potential at 0 V, and a dynode system to multiply the return beam. The photoelectrons are focused onto the glass target by means of a focusing electrode; the means for focusing the electron beam and guiding it toward the dynode system are not shown.

The photocathode is at about -400 V and the glass target is at about 0 V, so that the photoelectrons are accelerated toward the glass target and produce secondary electrons (multiplication factor $\delta_t > 1$). These secondary electrons are collected by the collector screen and thereby charge the glass target. If I_p is the primary current, the charging current flowing out of the glass target is $I_p(\delta_t - 1)$. The charge on the target is, therefore, an exact replica of the light distribution on the photocathode. The scanning beam periodically (30 times per second) brings the potential of the glass target back to 0 V, and hence the return beam is modulated in the rhythm of the charge distribution on the target.

The Image Orthicon

The signal-to-noise power ratio at the input is

$$\left(\frac{S}{N}\right)_{in} = \frac{I_p^2}{2eI_pB} = \frac{I_p}{2eB} \tag{13.1}$$

since I_p shows full-shot noise. Here B is the bandwidth.

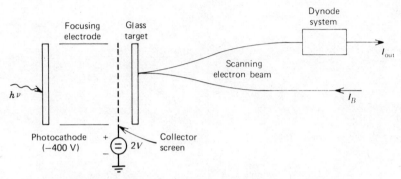

Figure 13.1. Schematic of an image orthicon (see text for details).

The glass target has a secondary emission factor δ_t and hence a gain $G_t = \delta_t - 1$. The noise of the target is

$$\overline{i_t^2} = 2eI_p B \cdot G_t^2 + 2eI_p B G_t P_t \tag{13.2}$$

Since $(\kappa_t - \delta_t)\delta_t$ is a measure for the secondary emission noise,

$$P_t = (\kappa_t - \delta_t)\delta_t / G(t) = (\kappa_t - \delta_t)\delta_t / (\delta_t - 1) \tag{13.2a}$$

The readout-beam current is I_B and the return current $I_R = I_B - G_t I_p$. The noise of the return beam is therefore

$$\overline{i_r^2} = \overline{i_{r0}^2} + \overline{i_t^2} = 2eI_R B + (G_t + P_t)G_t \cdot 2eI_p B \tag{13.3}$$

since the current I_R in and by itself shows full-shot noise:

$$\overline{i_{r0}^2} = 2eI_R B$$

If the dynode system has a gain G and a noise ratio P_m due to secondary emission noise,* the noise behind the dynode system is

$$\overline{i_a^2} = G^2\left(\overline{i_r^2} + 2eI_R P_m B\right)$$
$$= G^2\left[2eI_R B + (G_t + P_t)G_t \cdot 2eI_p B + 2eI_R P_m B\right] \quad (13.4)$$

The signal is contained in the part $G_t I_p$ of I_R, and hence the signal current is

$$I_a = GG_t I_p \quad (13.5)$$

The signal to noise power ratio at the output is thus

$$\left(\frac{S}{N}\right)_{out} = \frac{I_a^2}{\overline{i_a^2}} = \left(\frac{S}{N}\right)\frac{1}{F} = \frac{I_p}{2eB}\frac{1}{F} \quad (13.6)$$

where F is the noise-deterioration factor of the system

$$F = \frac{\overline{i_a^2}}{I_a^2} \cdot \frac{I_p}{2eB} = \frac{I_R + G_t(G_t + P_t)I_p + P_m I_R}{G_t^2 I_p} \quad (13.6a)$$

Substituting for I_R yields

$$F = \frac{I_B(1 + P_m) + G_t^2 I_p + G_t(P_t - P_m - 1)I_p}{G_t^2 I_p}$$

$$= 1 + \frac{P_t - P_m - 1}{G_t} + \frac{1 + P_m}{MG_t} \quad (13.7)$$

where

$$M = \frac{I_B - I_R}{I_B} = \frac{G_t I_p}{I_B} \quad (13.7a)$$

is the modulation depth of the readout beam.

*Since the targets may not all have the same κ and δ, P_m must be defined as

$$P_m = \sum_{j=1}^{n}\left[\frac{\kappa_j - \delta_j}{\prod\limits_{k=1}^{j}\delta_k}\right]$$

The Vidicon

Usually M is quite small, and the last term in (13.7) predominates. It comes about because of the velocity distribution of the electrons in the scanning beam and because the glass target has a high reflection coefficient for slow electrons.

13.2 THE VIDICON

In the vidicon the light shines through a transparent conducting electrode onto a photoconductor that thereby becomes conducting (Fig. 13.2). A voltage V of (ca. 10–20 V) is applied to the collecting electrode and this charges the surface of the photoconductor. That surface is scanned by an electron beam which restores the surface potential to 0 V 30 times per second.

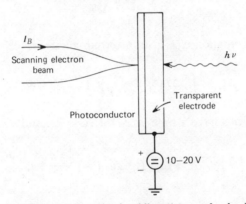

Figure 13.2. Schematic of a vidicon (see text for details).

Let the light produce N per second electrons (we assume that the holes are instantaneously trapped), then the photocurrent is, if L is the thickness of the photoconductor and μ the mobility of the carriers,

$$I = \frac{eV\mu}{L^2} \overline{N} \tag{13.8}$$

The charge Q transferred during the frame time $\tau_f(=1/30$ s) is therefore

$$Q = I\tau_f = \frac{eV\mu}{L^2} \overline{N} \tau_f \tag{13.8a}$$

We now assume that the scanning beam strikes the target for a time interval τ_d and that the beam fully discharges the target for a time $\tau_d' < \tau_d$. The charge transferred is $I_B \tau_d'$ and hence

$$I_B \tau_d' = I \tau_f, \quad \text{or} \quad \tau_d' = \frac{I}{I_B} \tau_f \tag{13.9}$$

Hence the current flowing in the external circuit during the discharge time τ_d' is

$$I_0 = \frac{I \tau_f}{\tau_d'} = I_B \tag{13.9a}$$

Next we turn to the noise. If N fluctuates by an amount ΔN, then the current fluctuation is

$$\Delta I = \frac{eV\mu}{L^2} \Delta N \tag{13.10}$$

and hence the fluctuation ΔQ_t in transferred charge is

$$\Delta Q_t = \int_0^{\tau_f} \Delta I(t) \, dt = \frac{eV\mu}{L^2} \int_0^{\tau_f} \Delta N(t) \, dt \tag{13.11}$$

If the photoelectrons have a time constant τ ($\tau \ll \tau_f$), and $s = v - u$,

$$\overline{\Delta Q_t^2} = \left(\frac{eV\mu}{L^2}\right)^2 \int_0^{\tau_f}\int_0^{\tau_f} \overline{\Delta N(u) \Delta N(v)} \, du \, dv = \left(\frac{eV\mu}{L^2}\right)^2 \tau_f \int_{-\infty}^{\infty} \overline{\Delta N^2} e^{-|s|/\tau} \, ds$$

$$= 2\left(\frac{eV\mu}{L^2}\right)^2 \overline{\Delta N^2} \tau_f \tau = \left(\frac{eV\mu}{L^2}\right)^2 \tau \overline{N} \tau_f \tag{13.12}$$

since $\overline{\Delta N^2} = \frac{1}{2}\overline{N}$, as is shown at the end of this section. Because of (13.9) this may be written

$$\overline{\Delta Q_t^2} = \frac{eV\mu}{L^2} I \tau \tau_f = \frac{eV\mu}{L^2} I_B \tau_d' \tau \tag{13.12a}$$

The fluctuation in the charge deposited by the beam is

$$\Delta Q_B = \int_0^{\tau_d'} \Delta I_B(t) \, dt = e \int_0^{\tau_d'} \Delta N_B(t) \, dt \tag{13.13}$$

where $\Delta N_B(t)$ is the fluctuation in the rate of arrival of electrons in the beam. Hence

$$\overline{\Delta Q_B^2} = e^2 \int_0^{\tau_d'} \int_0^{\tau_d'} \overline{\Delta N_B(u) \Delta N_B(v)}\, du\, dv = e^2\, \overline{\Delta N_B^2} \int_0^{\tau_d'} \int_0^{\tau_d'} \delta(u-v)\, du\, dv$$

$$= e^2\, \overline{N_B}\, \tau_d' = eI_B \tau_d' \tag{13.14}$$

since $\overline{\Delta N_B^2} = \overline{N_B}$, because the noise is shot noise.

Therefore, the total noise is

$$\overline{\Delta Q_{\text{tot}}^2} = \overline{\Delta Q_t^2} + \overline{\Delta Q_B^2} = eI_B \tau_d'\left[1 + \frac{V\mu\tau}{L^2}\right] = eI_B \tau_d'(1+G) \tag{13.15}$$

where $G = (\tau/\tau_{\text{dr}})$ is the photoconductor gain and $\tau_{\text{dr}} = (L^2/V\mu)$ is the drift time of the electrons.

The fluctuation ΔI_t in the output current thus has a mean square value

$$\overline{\Delta I_t^2} = \frac{\overline{\Delta Q_{\text{tot}}^2}}{\tau_d'^2} = \frac{eI_B}{\tau_d'}(1+G) = \frac{eI_B}{\tau_d'}\Gamma \tag{13.16}$$

where

$$\Gamma = 1 + G \tag{13.16a}$$

The reader will recognize eI_B/τ_d' as the shot noise in the beam for a time τ_d', whereas the factor Γ takes into account the effect of the target noise.

We finally prove the relation $\overline{\Delta N^2} = \tfrac{1}{2}\overline{N} = \tfrac{1}{2}N_0$, where N_0 is the equilibrium value of N. We know that $\overline{\Delta N^2} = g(N_0)\tau$. However, $g(N_0) = r(N_0) = \rho N_0^2 = Q$. Since $\tau = 1/(2\rho N_0)$, according to the definition of τ, we have $\overline{\Delta N^2} = Q\tau = \tfrac{1}{2}N_0$.

13.3 THE SECONDARY ELECTRON CONDUCTION VIDICON

Figure 13.3 shows a diagram of this type of vidicon. It shows resemblance to the vidicon in that the target is again scanned by an electron beam, and the charge deposited by the beam is read out at the output electrode.

The target consists of a thin Al_2O_3—Al layer onto which a porous KCl layer is deposited. Primary electrons of about 10 kV energy coming from a photocathode pass through the Al_2O_3 layer and the Al layer into the KCl layer, where they make a large number of secondary electrons that are collected by the wall screen of the vidicon gun. The primary beam is

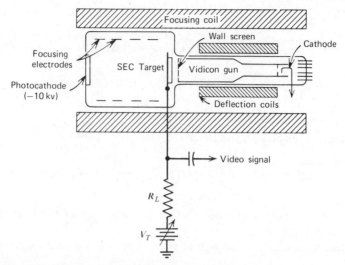

Figure 13.3. SEC vidicon schematic [according to G. W. Goetze and A. H. Boerio, *Proc. IEEE*, **52**, 1007 (1964); Aldert van der Ziel, *Solid State Physical Electronics*, 3rd Ed. © 1976, p. 272, Prentice Hall, Inc.].

therefore multiplied by the KCl layer by about a factor 200. As a consequence the multiplied primary noise is the chief source of noise.

Let I_{pr} be the primary current and I the secondary current. According to the variance theorem

$$I = I_{pr}\overline{a}\ ; \qquad S_I(f) = 2eI_{pr}\overline{a^2} \qquad (13.17)$$

where $\overline{a} = \delta$ is the secondary emission factor. Since $\Gamma = \overline{a^2}/(\overline{a})^2$ is the noise-deterioration factor, we may write

$$S_I(f) = 2eI_{pr}\delta^2\Gamma \qquad (13.17a)$$

This is the noise that is detected by the vidicon.

We must now correct this for the fact that not all incident electrons end up in the KCl. Let the part $1-\lambda$ be lost, mostly by back scattering. Then the current entering the KCl layer is λI_{pr} and hence

$$I = \lambda\overline{a}I_{pr} = \delta'I_{pr}; \qquad S_I(f) = 2e\lambda I_{pr}\delta^2\Gamma = 2eI_{pr}\delta'^2\Gamma' \qquad (13.18)$$

so that

$$\delta' = \lambda\delta; \qquad \Gamma' = \frac{\Gamma}{\lambda} \qquad (13.18a)$$

Here δ' is the *measured* multiplication factor and Γ' the measured noise-deterioration factor. For light targets such as Al—Al$_2$O$_3$ the factor λ should be relatively close to unity at high energies.

Timm* found Γ' to be relatively large for low primary energies (small δ', small λ), decreasing to about 1.5–2.0 for high primary energies (with δ' large, and λ close to unity). This is about what would be expected.

13.4 SOLID-STATE IMAGE SENSORS

A considerable effort has been made to replace the photoemissive and photoconductive targets by junction devices. The latest developments have led to photodiode arrays that can be scanned sequentially. For a good review of what can be achieved, see Weimer's review paper.[†]

13.4a Principle of Operation

Figure 13.4 shows a silicon photodiode target for a vidicon camera tube with electron-beam readout. The target consists of an array of approximately 10^6 photodiode elements diffused into a thin wafer of silicon, which is illuminated from the side opposite the diodes. During the integration time the p region of the diode collects holes that are subsequently discharged by the electron beam, resulting in a video output signal that can be amplified. It has light response both in the visible and in the near infrared.

Figure 13.5 shows an element of an MOS photodiode image sensor element. Here the source electrode of the device is insulated and acts as a photodiode collecting holes during the integration time. The gate is normally off, but periodically a scanning pulse is supplied that connects the source to the drain and that allows the hole charge stored in the source to move to the drain, resulting in a video-output pulse. It should be obvious that these sensor elements can be put into arrays and be scanned sequentially.

Figure 13.6 shows a third form using an array of MOS capacitors. Its principle is based on the fact that an MOS capacitor, produced by covering a silicon semiconductor chip with an insulating layer and a metal gate, can serve as an integrating photoelement similar to the diffused photodiode by biasing the gate so as to deplete the semiconductor surface. Illumination of the region causes the minority carriers to collect at the

---

*G. W. Timm and A. van der Ziel, *I.E.E.E. Trans.*, **ED-15**, 314 (1968).
[†]P. K. Weimer, Image Sensors for Solid State Cameras, in L. Marton (ed.), *Advances in Electronics and Electron Physics*, Vol. 37, Academic New York, pp. 181–262, 1975.

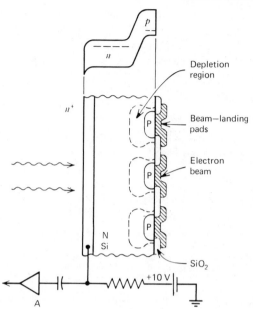

Figure 13.4. Cross section of a typical silicon photodiode target for a vidicon camera tube [P. K. Weimer, *Adv. Electron.*, **37** (1975)].

semiconductor–insulator interface. The accumulated charges can be removed laterally by charge transfer to an adjacent electrode or by charge injection back into the bulk semiconductor. The MOS capacitor photoelements are located on a thin slab of silicon and are illuminated through the silicon to avoid light interception by the metal gate.

Figure 13.7 shows multiplexed scanning of a single-line image sensor in which the elements comprise MOS photodetectors (*a*) or charge injection capacitor photodiodes (*b*). This can easily be extended to two-dimensional arrays.

The MOS capacitors used in Fig. 13.6 are known as *charge control devices* (CCD). In two-dimensional arrays of such devices one transfers charge from the one electrode to the next by sequential scanning. The most serious loss of signal is caused by failure to transfer the entire signal packet from one state to the next during a single clock cycle. A short signal pulse will accordingly be attenuated and shifted in phase as the residual charges are transferred during later clock cycles. A longer input pulse would reach full amplitude but with poor frequency response. When thousands of transfers must occur, for instance, and one wants normal television resolution, the fractional loss per transfer should not exceed 10^{-4} at a 10 MHz clock rate.

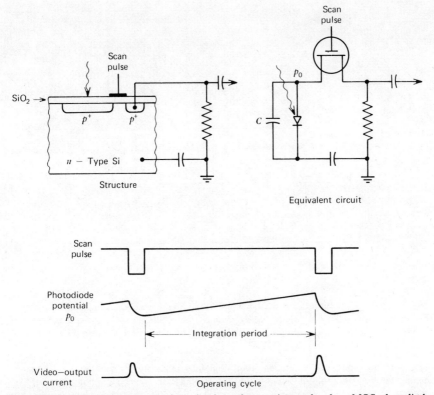

Figure 13.5. The structure, equivalent circuit, and operating cycle of an MOS photodiode image-sensor element [P. K. Weimer, *Adv. Electron.*, **37** (1975)].

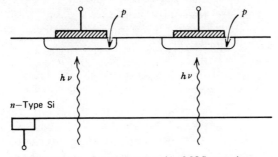

Figure 13.6. Image sensor using MOS capacitors.

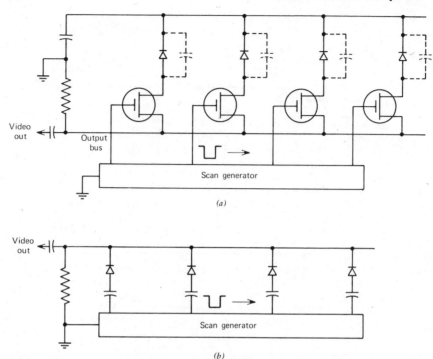

Figure 13.7. Multiplexed scanning of a single-line image sensor in which the elements comprise: (*a*) MOS photodetectors; (*b*) charge-injection capacitor photodiodes [P. K. Weimer, *Adv. Electron.*, **37** (1975)].

The most serious source of loss is trapping in the fast interface states of the CCD. To overcome this, one adds a residual charge to each CCD, a so-called "fat zero," so that most fast interface states are permanently filled. Figure 13.8 shows the resulting improvement in fractional loss per gate at lower clock frequencies.

13.4b Noise

We now discuss briefly the noise of these devices. A typical potential well, used during the charging time of the devices, holds about $N_s = 10^6$ carriers; the shot noise associated with this signal would correspond to $N_s^{1/2} = 10^3$ carriers, which is quite tolerable.

An MOS gated input circuit (see Figs. 13.5 and 13.6) can be used to introduce a fat-zero background. The inherent uncertainty in setting the fat-zero charge into the first well is set by thermal noise. Assuming the equipartition law to be valid, this would give an uncertainty

$$\left(\overline{N_n^2}\right)^{1/2} = \frac{1}{e}(kTC)^{1/2} = 400(C_{pF})^{1/2} \qquad (13.19)$$

where C is the capacitance of the MOS capacitor and C_{pF} is its value in picofarads.

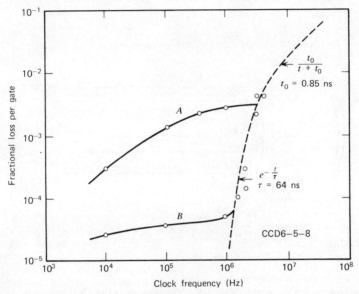

Figure 13.8. Effect of background current ("fat-zero") in reducing the transfer losses in a silicon gate two-phase surface-channel CCD register. Curve A had no "fat-zero" current and curve B had a "fat-zero" current of 50% of full well charge [P. K. Weimer, *Adv. Electron.*, **37** (1975)].

This estimate is probably somewhat too pessimistic, for we saw in Section 12.1c why the actual noise can be somewhat lower. Nevertheless it indicates that the effect of this noise source is comparable to that of the shot-noise source.

Another noise source is dark-current leakage. The magnitude of this effect depends strongly on temperature.

Weimer's review paper contains many references about these and similar problems.

13.5 THE PYROELECTRIC VIDICON

The pyroelectric vidicon* uses a poled pyroelectric target highlighted by infrared radiation and scanned by an electron beam (Fig. 13.9). There are two modes of operation: (a) the chopping mode and (b) the panning mode. In the chopping mode the incident radiation is chopped periodically and the camera is stationary; here the chopping produces a time-dependent temperature variation in, and hence a time-dependent voltage on, each target element. In the panning mode the incident radiation is not chopped but the camera is moved, thus producing a time-dependent voltage on each target element.

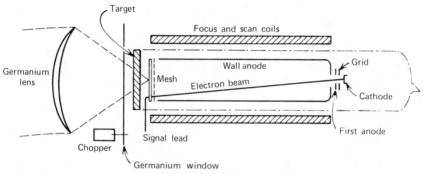

Figure 13.9. Pyroelectric vidicon [according to R. G. F. Taylor and H. A. H. Boot, *Contemp. Phys.*, **14**, 55 (1973)].

Since the target is nonconducting, the net current flowing to the target must be zero on the average. Nevertheless, one wants to provide beam readout. There are two types of readout; (a) anode stabilization and (b) cathode stabilization. In cathode-stabilization readout some residual gas is left in the tube, resulting in positive ions being collected at the target and making that target slightly positive. The electron beam just neutralizes this surface charge on the target and can read out both positive and negative signal voltages. In the anode-stabilization readout the target surface is positive so that it emits more secondary electrons than the primary electrons it receives. It will then stabilize at a voltage practically equal to the mesh voltage; again it can read both positive and negative signals.

The lateral conduction of heat in the target will reduce the magnitude of the local temperature changes and smear out the thermal image. This is

---

*R. G. F. Taylor and H. A. H. Boot, *Cont. Phys.*, **14**, 55 (1973).

measured by the modulation-transfer function (MTF), which represents the fractional reduction in signal current due to heat conduction. For the chopping mode*

$$(\text{MTF})_c = \left[1 + \left(\frac{2\pi k n^2}{f}\right)^2\right]^{-1/2} \tag{13.20}$$

and for the panning mode†

$$(\text{MTF})_p = \left[1 + \left(\frac{2\pi k n}{v}\right)^2\right]^{-1/2} \frac{\sin \pi n v t_f}{\pi n v t_f} \tag{13.21}$$

Here k is the thermal diffusivity in mm^2/s, n is the image resolution in line pairs/mm, f is the chopper frequency in Hz, v is the velocity in mm/s with which the image moves across the target, and $t_f = 16.6$ m/s. The term $[\sin(\pi n v t_f)]/(\pi n v t_f)$ is a panning readout efficiency term that is independent of the heat diffusivity.

The advantage of the pyroelectric vidicon is that the target does not need to be cooled and that it can operate in the 10–15 μm wavelength range, where room-temperature thermal radiation has its maximum intensity. Contrast is achieved either by variations in emissivity or by variations in temperature. Temperature differences down to a few tenths of a degree can be resolved in this manner.

---

*B. R. Holeman and W. M. Wreathall, *J. Phys. D: Appl. Phys.*, **4**, 1898 (1971).
†L. E. Garn and E. J. Sharp, *I.E.E.E. Trans.*, **PHP 10**, 208 (1974). This paper has a long list of references.

14

PHOTOMIXING

Lasers produce coherent light signals. For that reason signal processing techniques that were formerly used at radio and microwave frequencies have now become available for optical frequencies.

One such process is optical heterodyning or photomixing. It gives a large gain when compared to straight detection so that the noise of the post-amplifier is much less significant. Just as in a straight amplifier or heterodyne receiver (Section 14.1), its NEP is expressed in W/Hz, whereas in a straight detector P_{eq} is expressed in W/Hz$^{1/2}$ (Chapter 10). The photodiode, photovoltaic, and photoconductive mixers are discussed in some detail (Section 14.2). Section 14.3 discusses the point-contact mixer.

14.1 NOISE-EQUIVALENT POWER OF A HETERODYNE RECEIVER

Let a heterodyne receiver operating at microwave frequencies have a noise figure F. Then the noise at a frequency f can be represented by an equivalent input current generator

$$\overline{i_n^2} = F \cdot \frac{4kT\Delta f}{R_s} \tag{14.1}$$

as long as $(hf/kT) \ll 1$, and the corresponding equivalent available noise power is

$$P_{av} = \frac{1}{4}\overline{i_n^2} R_s = FkT\Delta f \tag{14.1a}$$

Consequently, the noise equivalent power is

$$P_{eq} = FkT \tag{14.2}$$

For $F=10$, $T=300°$ K, $P_{eq} = 4.1 \times 10^{-20}$ W/Hz.

Photomixing

This is much better than a straight detector, in which case we found $P_{eq} = 3 \times 10^{-13}$ W/Hz$^{1/2}$. One might, therefore, hope that optical mixing might give a similar improvement.

14.2 PHOTOMIXING

14.2a Gain

Suppose a light field $E_1 \cos \omega_1 t$ is to be detected. To that end this signal and a pump field $E_p \cos \omega_p t$ are made incident upon a photomultiplier, photodiode, photovoltaic cell, or photoconductive cell and the signal of difference frequency $\omega_0 = \omega_1 - \omega_p$ is filtered out. This signal can then be amplified by standard techniques.

To make the mixing work, the wavefronts of the incoming wave and the pump wave must be fully alligned, that is, the planes of constant phase in the two waves must be exactly parallel. This is not easy, but with some precaution it can be done. In addition, frequency control of the local oscillator is necessary to keep ω_0 constant.

Let the characteristic of the detector be

$$I = AE^2 \tag{14.3}$$

where A is a constant and $E = E_1 \cos \omega_1 t + E_p \cos \omega_p t$, then the total current is

$$A(E_1 \cos \omega_1 t + E_p \cos \omega_p t)^2$$
$$= AE_1^2 \cos^2 \omega_1 t + AE_p^2 \cos^2 \omega_p t + 2AE_1 E_p \cos \omega_1 t \cos \omega_p t$$
$$= \tfrac{1}{2} AE_1^2 + \tfrac{1}{2} AE_p^2 + AE_1 E_p \cos(\omega_1 - \omega_p) t + \text{harmonics} \tag{14.4}$$

If only the light signal $E_1 \cos \omega_1 t$ were present, the rectified current would be $I_1 = \tfrac{1}{2} AE_1^2$; if only the pump signal $E_p \cos \omega_p t$ were present the rectified current would be $I_p = \tfrac{1}{2} AE_p^2$. Due to the fact that *both* signals are present, there is a beat signal of frequency $\omega_1 - \omega_p$ with an amplitude

$$I_0 = AE_1 E_p = 2(I_1 I_p)^{1/2} \tag{14.5}$$

The power gain of the mixing process is therefore

$$G = \frac{\tfrac{1}{2} I_0^2}{I_1^2} = 2 \frac{I_p}{I_1} \tag{14.6}$$

where $\tfrac{1}{2} I_0^2$ represents the mean square of the beat signal.

At very low incoming signal levels the incoming signal drowns in the noise of the pump photons and (14.6) is no longer meaningful.

14.2b Noise in the Photodiode or Photomultiplier Photomixer

We first discuss the case of the photodiode. Let n be the rate of arrival of incoming photons and N the rate of release of photoelectrons. Then the quantum efficiency $\eta=(\overline{N}/\overline{n})$ and $I_1=e\overline{N}$ is the rectified current. Without pumping we would have

$$\text{var } N = \eta^2 \text{var } n + \overline{n}\,\eta(1-\eta) \tag{14.7}$$

If $\text{var } n = \overline{n}$, this becomes

$$\text{var } N = \overline{n}\,\eta = \overline{N} \tag{14.7a}$$

Now let N' be the number of electrons produced in the photomixing process. Since the amplification G works only on the first term in (14.7), the variance in N' for $G \gg 1$ is

$$\text{var } N' = G\eta^2 \text{var } n = G\eta^2 \overline{n} \tag{14.8}$$

so that the spectral intensity of the current fluctuation is

$$S_I'(f) = 2e^2 \text{var } N' = 2e^2 G\eta^2 \overline{n} = 2e\eta G I_1 = 4\eta e I_p \tag{14.9}$$

To this we must add the noise of the pump signal

$$S_I''(f) = 2eI_p \tag{14.10}$$

The total noise thus has a spectral intensity

$$S_I(f) = S_I'(f) + S_I''(f) = 2eI_p(1+2\eta) \tag{14.11}$$

To see the effect of the noise of the incoming signal we compare the total noise $S_I(f)$ to the noise $S_I''(f)$ of the pump only. Then

$$\frac{S_I(f)}{S_I''(f)} = 1 + 2\eta \tag{14.11a}$$

This was observed by Lee and van der Ziel for a silicon photodiode at 6328 Å.*

---

*S. J. Lee and A. van der Ziel, *Physica*, **45**, 379 (1969).

Photomixing

We finally discuss the signal-to-noise power ratio and the NEP, P_{eq}. If only the incoming signal were present, the noise in the straight detector would be $S_1(f) = 2eI_1$, and the signal-to-noise power ratio would be

$$\frac{S}{N} = \frac{I_1^2}{S(f)B} = \frac{I_1}{2eB} \tag{14.12}$$

where B is the bandwidth. For the photomixer we have instead

$$\frac{S}{N} = \frac{\frac{1}{2}I_0^2}{S_I(f)B} = \frac{2I_1 I_p}{2eI_p(1+2\eta)B} = \frac{2}{1+2\eta} \frac{I_1}{2eB} \tag{14.13}$$

so that the signal-to-noise ratio has changed by a factor $2/(1+2\eta)$, which decreases from the value 2 for $\eta \simeq 0$ to the value 2/3 for $\eta = 1$. The signal-to-noise ratio never changes very much in the mixing process. The advantage of the mixer is, however, that it increases the signal and the noise to such an extent that the noise of the postamplifier is much less significant.

If P_1 is the incident power, then $P_1/(eV_{ph})$ is the rate of incoming photons of frequency ν, where $eV_{ph} = h\nu$. Therefore $I_1 = e\eta P_1/(eV_{ph})$ and the signal-to-noise ratio for unit bandwidth is

$$\left(\frac{S}{N}\right) = \frac{2\eta}{1+2\eta} \cdot \frac{P_1}{2eV_{ph}}, \quad \text{or} \quad P_{eq} = eV_{ph} \frac{(1+2\eta)}{\eta} \tag{14.14}$$

As an example take $V_{ph} = 2.0$ eV and $\eta = 0.50$, then $P_{eq} = 1.3 \times 10^{-18}$ W/Hz.

This discussion holds for a solid-state photodiode and a photoemissive diode. For a photomultiplier the result is slightly different because the multiplication noise deteriorates the signal to noise ratio to a certain extent. For a photomultiplier with a gain δ^{2n} and a primary current I_{pr} the output noise has a spectrum

$$S_0(f) = 2eI_{pr}\delta^{2n}\Gamma = 2eI_{pr}\delta^{2n} + 2eI_{pr}\delta^{2n}(\Gamma - 1) \tag{14.15}$$

Here the first term is amplified shot noise of the primary current and the second term is the secondary emission noise, whereas $\Gamma = (\kappa - 1)/(\delta - 1)$ is the noise-deterioration factor. Bearing in mind (14.11) and taking into account that I_{pr} is equal to the pump current I_p, we thus find for a photomultiplier mixer for $G \gg 1$

$$S_I(f) = 2eI_p(1+2\eta)\delta^{2n} + 2eI_p\delta^{2n}(\Gamma - 1) = 2eI_p\delta^{2n}(\Gamma + 2\eta) \tag{14.16}$$

Consequently

$$P_{eq} = eV_{ph}\frac{\Gamma+2\eta}{\eta} \qquad (14.17)$$

For $\delta=5$ or $\Gamma=1.25$, $\eta=0.10$ and $V_{ph}=2.0\ eV$, this yields $P_{eq}=4.6\times10^{-18}$ W/Hz.

14.2c The Photovoltaic Photomixer

Here the expression for $S_I'(f)$ is still valid, but since in a photovoltaic cell two equal and opposite currents flow, each giving shot noise,

$$S_I''(f) = 4eI_p \qquad (14.18)$$

Consequently, for $G \gg 1$

$$S_I(f) = S_I'(f) + S_I''(f) = 2eI_p(2+2\eta) \qquad (14.19)$$

and

$$\frac{S}{N} = \frac{\frac{1}{2}I_0^2}{S_I(f)B} = \frac{2}{2+2\eta}\frac{I_1}{2eB} \qquad (14.20)$$

so that

$$P_{eq} = eV_{ph}\frac{2+2\eta}{\eta} \qquad (14.21)$$

The factor $2/(2+2\eta)$ varies between 1.0 for $\eta=0$ and $\frac{1}{2}$ for $\eta=1$. Putting $V_{ph}=2.0\ eV$, $\eta=0.50$ yields

$$P_{eq} = 1.9\times10^{-18}\ \text{W/Hz}$$

14.2d The Photoconductive Photomixer

Here we must change from currents to production rates. Let N' be the rate at which hole–electron pairs are produced by the incoming light and N'' the rate at which hole–electron pairs are produced by the pump light. Since the generation rate and the recombination rate add equal amounts of noise, we may write

$$\text{var}\,N'' = 2\overline{N''} \qquad (14.22)$$

Photomixing

According to (14.8) we have for the amplified photon noise for $G \gg 1$

$$\operatorname{var} N' = G\eta^2 \bar{n} = G\eta \overline{N_1'} \tag{14.23}$$

where $\eta = (\overline{N_1'}/\bar{n})$, $\bar{n}$ is the average rate of incoming photons, and $\overline{N_1'}$ is the average number of hole–electron pairs produced without mixing. Now G must be replaced by

$$G = 2 \frac{\overline{N''}}{\overline{N_1'}} \tag{14.24}$$

so that

$$\operatorname{var} N' = 2\eta \, \overline{N''} \tag{14.25}$$

Consequently

$$\operatorname{var} N = \operatorname{var} N' + \operatorname{var} N'' = (2 + 2\eta) \overline{N''} \tag{14.26}$$

so that the ratio of total noise to pump noise is

$$\frac{\operatorname{var} N}{\operatorname{var} N''} = \frac{2 + 2\eta}{2} \tag{14.26a}$$

just as for the photovoltaic cell.

Now we investigate the signal-to-noise power ratio. For the straight detector the average number of hole–electron pairs produced is $\overline{N_1'}$ and $S_{N_1}(0) = 2\overline{N_1'}$. Therefore, the input signal-to-noise ratio is

$$\left(\frac{S}{N}\right)_d = \frac{(\overline{N_1'})^2}{2 \overline{N_1'} B} = \frac{\overline{N_1'}}{2B} \tag{14.27}$$

In the mixing process the square of the signal is

$$G(\overline{N_1'})^2 = 2 \overline{N''} \, \overline{N_1'} \tag{14.28}$$

and the noise is

$$S_N(0) = 2 \operatorname{var} N = 2(2 + 2\eta) \overline{N''} \tag{14.29}$$

so that the signal-to-noise power ratio is

$$\left(\frac{S}{N}\right)_m = \frac{2 \overline{N''} \, \overline{N_1'}}{2(2 + 2\eta) N'' B} = \frac{2}{2 + 2\eta} \frac{\overline{N_1'}}{2B} = \frac{2}{(2 + 2\eta)} \left(\frac{S}{N}\right)_d \tag{14.30}$$

This is the same as for the photovoltaic cell, and hence for the photoconductive mixer

$$P_{eq} = eV_{ph}\frac{(2+2\eta)}{\eta} \quad (14.31)$$

Measurements by Lee agree with the theory.* For further details see references at the end of the chapter. Instead we found for the solid-state photodiode

$$P_{eq} = eV_{ph}\frac{1+2\eta}{\eta} \quad (14.31a)$$

Example. $\lambda = 10.6$ μm, or $V_{ph} = 0.117$ eV. Taking $\eta = 0.15$, then (14.31) yields $P_{eq} = 2.9 \times 10^{-19}$ W/Hz, which is quite good. This photomixer must, of course, be cooled.

14.3 THE POINT-CONTACT SCHOTTKY BARRIER DIODE MIXER

We saw in Section 10.5a how point-contact Schottky barrier diodes could be operated as detectors in the infrared. We shall now see how they can be used as photomixers in that frequency range.

The equivalent circuit of the point contact diode mixer is the same as the one shown in Fig. 10.9b for the detector. C is again the capacitance of the space-charge region, g_i the conductance of the space charge region, as modified by transit-time effects in the space-charge region, and r the series resistance of the bulk region of the diode.

When transit-time effects in the space-charge region must be taken into account, the local oscillator voltage of the mixer should be kept relatively small. For if the transit time of the carriers becomes approximately equal to the period of the local oscillator voltage, and the local oscillator voltage were large, practically no electrons would reach the point contact and hence no mixing would occur. Only at low barrier height and relatively small local oscillator voltage amplitudes v_p would one expect a considerable flow of d.c. and i.f. current. When that is the case, a linear approximation of the mixer operation becomes feasible. One can then make a Taylor expansion of the current I caused by a voltage V applied to the space-charge region. We have, if $V = (V_0 + \Delta V)$, and $e\Delta V/kT$ is relatively small,

$$I = I_{d0} + \frac{dI}{dV}\bigg|_{V_0} \Delta V + \frac{1}{2}\frac{d^2I}{dV^2}\bigg|_{V_0}\Delta V^2 + \cdots \quad (14.32)$$

*S. J. Lee and A. van der Ziel, *Physica*, **67**, 119 (1973).

The Point-Contact Schottky Barrier Diode Mixer

where $g_0 = (dI/dV)|_{V_0}$ is the low-frequency input conductance of the detector. If $\Delta V = (v_p \cos\omega_p t + v_i \cos\omega_i t)$, the current of frequency $|\omega_p - \omega_i|$ has an amplitude

$$\frac{1}{2} \frac{d^2 I}{dV^2}\bigg|_{V_0} v_p v_i \tag{14.33}$$

so that the low-frequency conversion conductance is

$$g_{0i0} = \frac{1}{2} g_0 \left(\frac{d^2 I/dV^2}{dI/dV}\right)_{V_0} v_p \tag{14.34}$$

From Fig. 10.9b and (10.54) replacing v_0 by v_i, we obtain

$$|v_i|^2 = \frac{8 P_a R_a}{[1 + g_i(R_a + r)]^2 + \omega^2 C^2 (R_a + r)^2} \tag{14.35}$$

where $P_a = |v_a|^2 / 8 R_a$.

We must now bear in mind that the mixing process has the same transit-time deterioration as the detection process; we thus have at the highest frequencies a conversion admittance Y_{0i} given by

$$Y_{0i} = g_{0i0} |g(j\omega)|^2 \tag{14.36}$$

The space-charge region has a low-frequency conductance $g_0 = (dI/dV)_{V_0}$. If $r \ll (1/g_0)$, we can neglect the effect of the series resistance r in the output. The available output signal power is then

$$(P_{av})_{out} = \frac{|Y_{0i}|^2 |v_i|^2}{8 g_0} = \frac{1}{8}\left(\frac{g_{0i0}}{g_0}\right)^2 |g(j\omega)|^4 g_0 |v_i|^2 \tag{14.37}$$

so that the available power gain of the mixer is

$$G_{av} = \frac{(P_{av})_{out}}{P_a} = \left(\frac{g_{0i0}}{g_0}\right)^2 |g(j\omega)|^4 \frac{g_0 R_a}{[1 + g_i(R_a + r)]^2 + \omega^2 C^2 (R_a + r)^2} \tag{14.38}$$

Now the short-circuit output noise in a frequency interval Δf has a mean square value

$$\overline{i_0^2} = 2 e I_{d0} \Delta f = n_0 \cdot 4 k T g_0 \Delta f, \quad \text{where} \quad n_0 = \frac{e I_{d0}}{2 k T g_0} \tag{14.39}$$

is the noise ratio of the mixer output. The i.f. amplifier noise can be represented by a current generator

$$\overline{i_a^2} = (F_2 - 1)4kTg_0\Delta f \tag{14.39a}$$

in parallel with g_0, where F_2 is the noise figure of the i.f. amplifier for the source resistance $1/g_0$. Hence the total noise can be represented by a current generator

$$\overline{i_{\text{tot}}^2} = (n_0 + F_2 - 1)4kTg_0\Delta f \tag{14.40}$$

in parallel with g_0. Consequently, the equivalent available noise power per unit bandwidth at the output is

$$(P_{\text{eq}})_0 = \frac{\overline{i_{\text{tot}}^2}}{4g_0\Delta f} = (n_0 + F_2 - 1)kT \tag{14.40a}$$

and hence the equivalent noise power at the antenna is

$$(P_{\text{eq}})_a = \frac{n_0 + F_2 - 1}{G_{\text{av}}}kT \tag{14.41}$$

To make $(P_{\text{eq}})_a$ as small as possible, one should make G_{av} as large as possible. This means that the transit-time deterioration factor $|g(j\omega)|^2$ should be made as close to unity as possible and the resistivity ρ of the bulk material as small as possible; both can be achieved by high doping. As in the detector, it pays to make the contact area A as small as feasible, e.g. $A = 10^{-11}$ cm^2, but one should not go so far as to make $r > R_a$.

Up to now we treated the case of linear mixing in which G_{av} varies as the square of the local oscillator amplitude v_p. Since the mixing deteriorates at very high local oscillator amplitudes, there should be a value of v_p for which G_{av} has its optimum value. This optimization problem is difficult to solve theoretically, but is easily solved experimentally since one can adjust v_p for maximum i.f. output.

The best results are again expected for the n-type GaAs point-contact detector. Good mixing down to 10 μm wavelength should be feasible for doping levels of the order of $10^{19}/$cm^3.

One final remark: the available power P_a of the antenna is generally smaller than the available incident power P_i. Putting $P_a = aP_i$, we should thus have $a < 1$ and hence the equivalent input noise power in the incident beam is

$$(P_{\text{eq}})_i = \frac{(P_{\text{eq}})_a}{a} \tag{14.42}$$

REFERENCES

R. K. Willardson and A. C. Beer, (eds.), *Semiconductors and Semimetals*, Vol. 5, Academic, New York (1970): see Chapter 8, R. J. Keyes and T. M. Quist, Low-level coherent and incoherent detection in the infrared; Chapter 9, M. C. Teich, Coherent detection in the infrared; Chapter 10, F. R. Arams, E. W. Sard, B. J. Peyton, and F. P. Pace, Infrared heterodyne detection with gigahertz IF response.

A. van der Ziel, *J. Appl. Phys.*, **47**, 2059 (1976) (Section 14.3).

15

LIGHT AMPLIFICATION WITH CATHODELUMINESCENCE

An interesting light-amplification scheme, based on cathodeluminescence, will be discussed in this chapter. Here a photocathode is illuminated by a weakly illuminated scene focused onto it. The photocathode then emits photoelectrons that are accelerated in a potential difference of about 20 kV and focused on a luminescent screen. The luminescent light output of the screen is then larger than the light input at the photocathode.

To understand the noise behavior of such a light amplifier, one must know the noise generated in the luminescence process. This is done in Section 15.1, whereas Section 15.2 discusses the applications.

15.1 NOISE IN CATHODELUMINESCENT LIGHT*

Let n_p be the number of electrons arriving at the luminescent material per second and let the incoming electrons show full-shot noise, that is, $\operatorname{var} n_p = \overline{n_p}$. Let the ith electron produce P_i photons so that

$$N = \sum_{i}^{n_p} P_i \tag{15.1}$$

photons are produced in a given second. Hence if $\overline{P_i} = P$ and $\operatorname{var} P_i = \operatorname{var} P$ for all i, the variance theorem predicts

$$\overline{N} = \overline{n_p}\,\overline{P}\,; \quad \operatorname{var} N = \left(\overline{P}\right)^2 \operatorname{var} n_p + \overline{n_p} \operatorname{var} P = \overline{n_p}\,\overline{P^2} \tag{15.2}$$

---

*T. M. Chen and A. van der Ziel, *I.E.E.E. Trans.*, **ED-12**, 489 (1965).

If there were no noise generated in the luminescence we would have

$$\operatorname{var} N = \overline{n_p} (\overline{P})^2 \qquad (15.2a)$$

We thus write

$$\operatorname{var} N = \overline{n_p} (\overline{P})^2 \Gamma, \quad \text{where} \quad \Gamma = \frac{\overline{P^2}}{(\overline{P})^2} \qquad (15.2b)$$

is the noise-deterioration factor. Hence the low-frequency noise is

$$S_N(0) = 2 \operatorname{var} N = 2 \overline{n_p} (\overline{P})^2 \Gamma \qquad (15.3)$$

We now evaluate the frequency dependence of the photon noise. To that end we must specify the model of the luminescent material more closely. We consider a noncharacteristic luminescent material (like ZnS) as follows. The incoming electrons generate hole–electron pairs; the holes are immediately trapped and the free electrons recombine with the trapped holes under the emission of of light. Losses due to nonemitting transitions via recombination centers are neglected. The electrons and holes thus consist of equal numbers, denoted by n. The rate equation for the generation–recombination process is

$$\frac{dn}{dt} = g_0 - Cn^2 + g(t) - r(t) \qquad (15.4)$$

where g_0 is the net generation rate of hole–electron pairs and $r(n) = Cn^2$ is the recombination rate; here g_0 and C are constants. Finally $g(t)$ and $r(t)$ are random-source functions describing the fluctuations in the generation and recombination rates, respectively. The equilibrium number of free electrons, n_0, follows from

$$g_0 = Cn_0^2 \qquad (15.4a)$$

and $g(t)$ and $r(t)$ have spectra

$$S_g(f) = 2 \operatorname{var} N = 2 \overline{n_p} (\overline{P})^2 \Gamma; \quad S_r(f) = 2r(n_0) = 2\overline{N} = 2 \overline{n_p}\, \overline{P} \qquad (15.4b)$$

Looking for small fluctuations Δn in n around n_0 we find from a Taylor expansion of (15.4) around n_0,

$$\frac{d\Delta n}{dt} = -\frac{\Delta n}{\tau} + g(t) - r(t) \qquad (15.5)$$

where

$$\frac{1}{\tau} = \left[\frac{d(Cn^2)}{dn} - \frac{dg_0}{dn}\right]_{n_0} = 2Cn_0 \tag{15.5a}$$

We now substitute for $0 \leq t \leq T$

$$\Delta n = \sum_n a_n \exp(j\omega_n t); \quad g(t) = \sum_n b_n \exp(j\omega_n t); \quad r(t) = \sum_n c_n \exp(j\omega_n t)$$

and obtain

$$a_n = \frac{(b_n - c_n)\tau}{1 + j\omega\tau} \tag{15.6}$$

The fluctuation in light emission is in $(\Delta n/\tau) + r(t)$, which has a Fourier coefficient

$$\frac{a_n}{\tau} + c_n = \frac{b_n}{1 + j\omega\tau} + c_n \frac{j\omega\tau}{1 + j\omega\tau} \tag{15.6a}$$

so that

$$S_N(f) = \frac{S_g(f)}{1+\omega^2\tau^2} + \frac{S_r(f)\omega^2\tau^2}{1+\omega^2\tau^2} = \frac{2\overline{N}\,\overline{P}\Gamma}{1+\omega^2\tau^2} + 2\overline{N}\frac{\omega^2\tau^2}{1+\omega^2\tau^2} \tag{15.7}$$

where $\overline{N} = \overline{n_p P}$ as before. At high frequencies $(\omega\tau \gg 1)$ one thus obtains full-shot noise.

We now rewrite $\overline{P}$ by introducing the energy efficiency of the luminescence. Since eV_a is the primary energy available and $eV_{ph} = h\nu$, the maximum possible number of photons is (V_a/V_{ph}). Therefore, the efficiency η_c is defined by the relation

$$\overline{P} = \eta_c \frac{V_a}{V_{ph}} \tag{15.8}$$

Typically $\eta_c = 10$–20%. For $V_a = 25$ kV, $V_{ph} = 2.5$ eV, and $\eta_c = 0.20$, on the average 2000 photons are generated per primary electron. Hence the noise given by (15.7) is strongly super-Poissonian.

However, this is not what is *measured*. To measure the noise in the light emission, the light has to be detected by a photoemissive diode. Let η_{ph} be the probability that a produced photon is collected by the photocathode of the diode, and let η_m be the quantum efficiency of the photocathode, so that $\eta_{ph}\eta_m$ is the probability that a produced photon makes a photoelectron. Therefore if m photoelectrons are produced per second, the variance

Noise in Cathodeluminescent Light

theorem yields

$$m = \overline{N} \eta_{ph}\eta_m = \overline{n_p} \, \overline{P} \eta_{ph}\eta_m \tag{15.9}$$

$$\operatorname{var} m = (\eta_{ph}\eta_m)^2 \operatorname{var} N + \overline{N}\eta_{ph}\eta_m(1 - \eta_{ph}\eta_m) \tag{15.10}$$

Therefore, at low frequencies

$$S_m(0) = 2 \operatorname{var} m = (\eta_{ph}\eta_m)^2 S_N(0) + 2\overline{N}\eta_{ph}\eta_m(1 - \eta_{ph}\eta_m)$$

$$= 2\overline{n_p}\left[(\overline{P}\eta_{ph}\eta_m)^2 \Gamma + \overline{P}\eta_{ph}\eta_m(1 - \eta_{ph}\eta_m)\right] \tag{15.11}$$

Now

$$M = \overline{P}\eta_{ph}\eta_m = \frac{V_a}{V_{ph}}\eta_c\eta_{ph}\eta_m \tag{15.12}$$

is the multiplication factor of this arrangement. If $V_a = 25$ kV, $V_{ph} = 2.5$ eV, $\eta_c = 0.20$, $\eta_{ph} = 0.50$, and $\eta_m = 0.10$, then $M = 100$. This is a large number. Hence the second term in (15.11) may be neglected and

$$\overline{m} = \overline{n_p} M; \quad S_m(0) \simeq 2\overline{n_p} M^2 \Gamma \tag{15.13}$$

and the high-frequency noise is

$$S_m(f) = \frac{2\overline{n_p} M^2 \Gamma}{1 + \omega^2 \tau^2} + 2\overline{n_p} M(1 - \eta_{ph}\eta_m) + 2\overline{n_p} M\eta_{ph}\eta_m \frac{\omega^2 \tau^2}{1 + \omega^2 \tau^2} \tag{15.13a}$$

since the last term in (15.7) is multiplied by $\eta_{ph}^2\eta_m^2$, and $M = \overline{P}\eta_{ph}\eta_m$.

We now switch to currents $I_{pr} = e\overline{n_p}$ and $I_0 = e\overline{m} = eM\overline{n_p} = MI_{pr}$, and we obtain

$$S_I(0) \simeq 2eI_0 M\Gamma \tag{15.14}$$

$$S_I(f) = \frac{2eI_0 M\Gamma}{1 + \omega^2 \tau^2} + 2eI_0\left[1 - \frac{\eta_{ph}\eta_m}{1 + \omega^2 \tau^2}\right] \tag{15.14a}$$

Since M is a large number, the last term is negligible except for the highest frequencies where the noise is full-shot noise. In the next section we shall, therefore, neglect these terms. Since I_0 and $M = (I_0/I_{pr})$ are measured and $S_I(0)$ and τ can be determined from the measurements, Γ can be evaluated.

Most phosphors are a mixture of materials with different time constants and hence the spectrum should be somewhat "smeared out." This is found experimentally.

It is not difficult to extend the theory to include losses in the photon production or to evaluate the case of characteristic luminescence.*

15.2 NOISE IN LIGHT AMPLIFIERS

In this case the photocathode comes first. Let it have a quantum efficiency η_m, let the energy efficiency of the luminescent material be η_c and let the part η_{ph} of the light be collected. Going through the same reasoning as before, the gain M of the light amplifier is found to be

$$M = \eta_m \eta_c \eta_{ph} \frac{V_a}{V_{ph}} \qquad (15.15)$$

For $\eta_m = 10\%$, $\eta_c = 20\%$, $\eta_{ph} = 50\%$, $V_a = 25$ kV, $V_{ph} = 2.5$ eV, we have $M = 100$, as before.

Much higher gains could have been obtained by using two amplifier stages in cascade and applying 12.5 kV to each stage. The gain of each stage would have been 50 and the total gain 2500. This is certainly a worthwhile improvement.

We now calculate the noise of a two-stage amplifier in two steps. At the end of the third step we then have the noise of a single stage amplifier.

1. *Primary quanta*: $\overline{K}$ per second

$$S_a(0) = 2\overline{K} \qquad (15.16)$$

2. *Primary electrons*: $\eta_m \overline{K}$ per second

$$S_b(0) = 2\overline{K}\eta_m \qquad (15.17)$$

3. *Secondary quanta*:

$$\overline{N} = \overline{K}\eta_m \overline{P}\eta_{ph} = \overline{K}M \qquad (15.18)$$

per second where $M = \eta_m \eta_{ph}\overline{P} = (\eta_m \eta_{ph} \eta_c V_a / V_{ph})$ is the multiplication factor,

$$S_c(0) = 2\overline{K}\eta_m (\overline{P}\eta_{ph})^2 \Gamma = \frac{2\overline{K}}{\eta_m} M^2 \Gamma \qquad (15.19)$$

*R. J. J. Zijlstra, *Proc. Int. Conf. Lumin.*, Vols. 1 and 2, Budapest, Academia Krado (1968), pp. 2102–2107; H. M. Fijnaut and R. J. J. Zijlstra, *Brit. J. Appl. Phys.*, **3**, 45 (1970).

Noise in Light Amplifiers

and, if we neglect a few small terms

$$S_c(f) = \frac{2\overline{K}}{\eta_m} \frac{M^2 \Gamma}{(1+\omega^2\tau^2)} \qquad (15.19a)$$

where τ is the time constant of the luminescent screen. This represents the noise of a single stage.

For the remainder of the calculation we write

$$S_c(0) = 2\overline{K}\eta_m(\overline{P}\eta_{ph})^2 + 2\overline{K}(\overline{P}\eta_{ph})^2\eta_m(\Gamma-1) \qquad (15.19b)$$

where the first term is the *amplified shot noise* and the second term is the *true luminescent noise*.

4. *Secondary electrons*: Per second $\overline{K}\eta_m^2\overline{P}\eta_{ph} = \overline{K}\eta_m M$

$$S_d(0) = \eta_m^2 S_c(0) = 2\overline{K}\eta_m(\overline{P}\eta_m\eta_{ph})^2 \Gamma = 2\overline{K}\eta_m M^2 \Gamma \qquad (15.20)$$

5. *Tertiary quanta*: Per second

$$\overline{K}\eta_m^2(\overline{P}\eta_{ph})^2 = \overline{K}M^2 \qquad (15.21)$$

$$S_e(0) = 2\overline{K}\eta_m(\overline{P}\eta_m\eta_{ph})^2 \cdot (\overline{P}\eta_{ph})^2 \Gamma + 2\overline{K}\eta_m^2\overline{P}\eta_{ph} \cdot (\overline{P}\eta_{ph})^2(\Gamma-1) \qquad (15.22)$$

where $\overline{K}\eta_m^2\overline{P}\eta_{ph} = \overline{K}\eta_m M$ is the rate of arrival of electrons at the second screen. The shot noise of this number is multiplied by $(\overline{P}\eta_{ph})^2(\Gamma-1)$ [see (15.19b)].

Introducing M, this may be written

$$S_e(0) = \frac{2\overline{K}}{\eta_m} M^4 \left[\Gamma + \frac{\Gamma - 1}{M} \right] \qquad (15.23)$$

The second term is small if M is large. We may then write, if we neglect a few other small terms and take into account the frequency dependence of both luminescent screens

$$S_e(f) = \frac{2(\overline{K}/\eta_m)M^4\Gamma}{(1+\omega^2\tau^2)^2} \qquad (15.24)$$

Timm measured Γ as a function of M and found that Γ was rather large for small M (i.e., small accelerating voltage V_a) but decreased strongly with increasing V_a, reaching a limiting value of about 1.5–2.0 for relatively large values of V.* This is not an unreasonable result.

*G. W. Timm and A. van der Ziel, *I.E.E.E. Trans.*, **ED-15**, 314 (1968).

16

JOSEPHSON JUNCTION DEVICES

A Josephson junction consists of two superconducting materials separated by a thin oxide layer. The simplest configuration is to make a point contact between a pointed superconducting wire and a superconducting plate.

In a superconductor the electrons can exist in two forms: (a) as normal electrons that are scattered by the lattice in the normal manner and (b) as Cooper pairs that cannot be scattered by the lattice and that are responsible for the superconductive properties of the material.

The Cooper pairs and the normal electrons can pass through the oxide layer by tunnel effect. This has two effects:

1. A direct supercurrent can flow up to some maximum value I_0 without any voltage being developed across the sample. This is called the *d.c. Josephson effect*.

2. If a d.c. bias voltage V_0 is applied to the junction, the paired electron current alternates with a frequency $f_0 = (2eV_0/h)$, where e is the electron charge and h is Planck's constant; the device thus becomes a single-frequency oscillator. For $V_0 = 1 \ \mu V$ we have $f_0 = 483.6$ MHz. This is called the *a.c. Josephson effect*. For a proof of this property see Section 16.2b.

We further need a superconducting property called *flux quantization*, according to which the flux ϕ in a superconducting ring can only have values $s\phi_0$; where $s = 0, 1, 2 \ldots$ and $\phi_0 = (h/2e)(= 2.07 \times 10^{-15}$ weber). This is a macroscopic quantum effect.

Finally we need an expression for the noise in a Josephson junction. We shall show that the current in a Josephson junction has a spectral intensity

$$S_I(f) = 4kT\frac{I}{V} \tag{16.1}$$

where I is the current passed through the device and V is the applied voltage. It should be noted that at $V=0$ the current changes from $-I_0$ to $+I_0$ so that the device characteristic has a kink at $V=0$.

To prove (16.1) we denote the normal electrons by n and the Cooper pairs by p. The normal electrons give rise to a normal current $I_n = I_{n1} - I_{n2}$, where I_{n1} and I_{n2} are two normal currents that flow across the junction in opposite directions. The paired electrons give rise to a pair current $I_p = I_{p1} - I_{p2}$, where I_{p1} and I_{p2} are two pair currents that flow across the junction in opposite directions. Hence

$$I = I_n + I_p = I_{n1} - I_{n2} + I_{p1} - I_{p2} \tag{16.1a}$$

All currents fluctuate independently and each shows full-shot noise. However, since the pairs carry twice as much charge as singles

$$S_{I_n}(f) = 2e(I_{n1} + I_{n2}) = 2eI_n \frac{I_{n1} + I_{n2}}{I_{n1} - I_{n2}} \tag{16.1b}$$

$$S_{I_p}(f) = 4e(I_{p1} + I_{p2}) = 4eI_p \frac{I_{p1} + I_{p2}}{I_{p1} - I_{p2}} \tag{16.1c}$$

A wave-mechanical calculation shows that [cf. (5.25a)]

$$\frac{I_{n1} + I_{n2}}{I_{n1} - I_{n2}} = \coth\left(\frac{eV}{2kT}\right); \quad \frac{I_{p1} + I_{p2}}{I_{p1} - I_{p2}} = \coth\left(\frac{2eV}{2kT}\right); \tag{16.1d}$$

the latter expression takes into account that the pairs carry a charge $2e$. Consequently

$$S_I(f) = S_{I_n}(f) + S_{I_p}(f) = 2eI_n \coth\left(\frac{eV}{2kT}\right) + 4eI_p \coth\left(\frac{eV}{kT}\right) \tag{16.1e}$$

For small values of $|eV/kT|$ we have $\coth(eV/2kT) \simeq 2kT/eV$ and $\coth(eV/kT) \simeq kT/eV$. Hence for small $|V|$

$$S_I(f) = 2eI_n\left(\frac{2kT}{eV}\right) + 4eI_p\left(\frac{kT}{eV}\right) = 4kT\frac{(I_n + I_p)}{V} = 4kT\frac{I}{V}$$

as had to be proved.

16.1 THE JOSEPHSON JUNCTION AS A THERMOMETER

If a d.c. voltage V_0 only were applied to the Josephson junction, the device would be an oscillator generating a single frequency $f_0 = (2eV_0/h)$. As a result of the inherent noise present in the circuit, a noise voltage $v(t)$ is developed across the junction and this produces an f.m. noise-modulated signal of frequency

$$f = f_0 + \Delta f(t) = f_0 + \frac{2ev(t)}{h} \tag{16.2}$$

since the applied voltage is now $V_0 + v(t)$. As a consequence the oscillator has a certain linewidth. A calculation shows that the linewidth is directly proportional to the temperature T. Having evaluated the circuit parameters, one can uese linewidth measurements of the Josephson junction oscillator as a means of determining the temperature of the system. Such Josephson junction thermometers for the m° K range are now commercially available.

16.1a Discussion of the Josephson Thermometer

We saw from (16.2) that

$$\Delta f(t) = \frac{2ev(t)}{h} \quad \text{or} \quad S_{\Delta f}(f) = \left(\frac{2e}{h}\right)^2 S_v(f) \tag{16.2a}$$

where $S_{\Delta f}(f)$ is the spectral intensity of $\Delta f(t)$.

Now we switch notation slightly by introducing spectral intensities that are half as large as the values normally used in this book. We do so to conform with the theory of f.m. noise modulated signals. $S_x(f)$ then means the spectrum for a positive or negative frequency f; whereas in our previous notation $S'_x(f) = S_x(f) + S_x(-f) = 2S_x(f)$, since $S_x(f)$ is symmetric in f.

We now turn to the equivalent circuit of Fig. 16.1. It consists of a Josephson junction J shunted by a bias resistance r of the order of 1 mΩ or less. The two current generators $(2kT\Delta f/r)^{1/2}$ and $(2kT\Delta fI/V_0)^{1/2}$ represent, respectively, the thermal noise of the resistance r and the noise of a Josephson junction carrying a current I when a voltage V_0 is applied to the junction. Since the differential resistance of J is large in comparison with

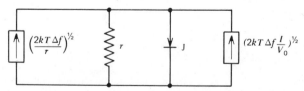

Figure 16.1. Equivalent circuit of a Josephson oscillator used for temperature measurements.

The Josephson Junction as a Thermometer

the bias resistance r, we have

$$S_v(f) = \left(\frac{2kT}{r} + 2kT\frac{I}{V_0}\right)r^2 = 2kTr\left(1 + \frac{Ir}{V_0}\right) \quad (16.3)$$

Consequently

$$S_{\Delta f}(f) = \left(\frac{2e}{h}\right)^2 \cdot 2kTr\left(1 + \frac{Ir}{V_0}\right) \quad (16.4)$$

We are now interested in the spectrum of the random variable

$$x(t) = \exp\{j\omega_0 t + j[2\pi\Delta f(t)]t\} \quad (16.5)$$

Calculation shows that

$$S_x(f) = \frac{1}{\pi^2 S_{\Delta f}(0)} \frac{1}{1 + \{(f-f_0)/[\pi S_{\Delta f}(0)]\}^2} \quad (16.6)$$

so that the width Δf_0 between half-power points is

$$\Delta f_0 = 2\pi S_{\Delta f}(0)$$

Consequently, the line width Δf_0 of the f.m. modulated Josephson junction signal is

$$\Delta f_0 = \pi \left(\frac{2e}{h}\right)^2 \cdot 4kTr\left(1 + \frac{Ir}{V_0}\right) \quad (16.7)$$

Measuring Δf_0 and knowing r, I, and V_0 we can thus evaluate T. To measure Δf_0 one must use a narrow-band receiver with a bandwidth $B \ll \Delta f_0$.

16.1b Proof of (16.6)

To prove (16.6) we rewrite

$$x(t) = \exp[j\omega_0 t + j\psi(t)] = \exp(j\omega_0 t)u(t) \quad (16.8)$$

and define

$$\Delta f(t) = \frac{1}{2\pi}\frac{d\psi(t)}{dt}, \quad \text{or} \quad S_{\Delta f}(f) = f^2 S_\psi(f) \quad (16.9)$$

so that

$$S_\psi(f) = \frac{S_{\Delta f}(f)}{f^2} \tag{16.9a}$$

Since we know $S_{\Delta f}(f)$, we also know $S_\psi(f)$.

We now want the spectrum of

$$u(t) = \exp[j\psi(t)] \tag{16.10}$$

To evaluate $S_u(f)$ we must know the autocorrelation function

$$\varphi_u(s) = \overline{u^*(t)u(t+s)} = \overline{\exp[-j\psi(t)+j\psi(t+s)]} \tag{16.11}$$

where the asterisk denotes the complex conjugate. To evaluate this, we must know the joint probability density function $W(\psi_1,\psi_2)$ of $\psi_1 = \psi(t)$ and $\psi_2 = \psi(t+s)$. According to Chapter 4, by evaluating (4.9), we obtain

$$W(\psi_1,\psi_2) = \frac{1}{2\pi[\phi_\psi^2(0)-\phi_\psi^2(s)]^{1/2}} \exp\left[\frac{-\tfrac{1}{2}\phi_\psi(0)(\psi_1^2+\psi_2^2)+\phi_\psi(s)\psi_1\psi_2}{\phi_\psi^2(0)-\phi_\psi^2(s)}\right] \tag{16.12}$$

where $\phi_\psi(s) = \overline{\psi^*(0)\psi(0+s)}$ and $\phi_\psi(0) = \overline{\psi^*(0)\psi(0)}$ and the asterisk takes into account the possibility that $\psi(t)$ is complex; we have here taken $t=0$. Then, after some manipulations

$$\phi_u(s) = \int_{-\infty}^{\infty}\int_{-\infty}^{\infty} \exp(-j\psi_1+j\psi_2) W(\psi_1,\psi_2)\, d\psi_1\, d\psi_2$$

$$= \exp[-\phi_\psi(0)+\phi_\psi(s)] = \exp[-k(s)] \tag{16.13}$$

Now by inversion

$$\phi_\psi(s) = \int_{-\infty}^{\infty} S_\psi(f)\exp(2\pi jfs)\, df$$

and hence

$$k(s) = \phi_\psi(0) - \phi_\psi(s) = \int_{-\infty}^{\infty} S_\psi(f)[1-\exp(2\pi jfs)]\, df$$

$$= 2\int_{-\infty}^{\infty} S_\psi(f)(\sin \pi fs)^2\, df$$

$$= 2\pi s S_{\Delta f}(0)\int_{-\infty}^{\infty} \left(\frac{\sin \pi fs}{\pi fs}\right)^2 d(\pi fs) = 2\pi^2 s S_{\Delta f}(0) \tag{16.14}$$

since $S_{\Delta f}(f)$ has practically a white spectrum. Therefore

$$S_u(f) = \int_{-\infty}^{\infty} \phi_u(s)\exp(2\pi jfs)\,ds = \frac{4\pi^2 S_{\Delta f}(0)}{[2\pi^2 S_{\Delta f}(0)]^2 + (2\pi f)^2}$$

$$= \frac{1}{\pi^2 S_{\Delta f}(0)\left\{1 + \left[\dfrac{f}{\pi S_{\Delta f}(0)}\right]^2\right\}} \quad (16.15)$$

and

$$S_x(f) = S_u(f - f_0) = \frac{1}{\pi^2 S_{\Delta f}(0)\left\{1 + \left[\dfrac{f - f_0}{\pi S_{\Delta f}(0)}\right]^2\right\}}$$

as had to be proved.

16.2 THE JOSEPHSON JUNCTION AS AN AMPLIFIER

16.2a The Low-Frequency Josephson Amplifier

By applying a d.c. signal V_0 and a small a.c. signal $v_s \cos\omega_s t$ to a Josephson junction, one obtains an f.m. modulated signal of carrier frequency $f_0 = (2eV_0/h)$. This f.m. modulated signal can be received by an amplifier and detected by an f.m. detector to give an amplified replica of the a.c. signal of frequency ω_s.

To understand the bandwidth and noise limitation of this amplifier we take into account the inductance L of the input circuit. Typically L is of the order of 10^{-9} henry, but by proper design of the junction it can be made about a factor 10 smaller (Fig. 16.2a).

As far as the signal is concerned there is no limitation as long as ωL is small in comparison with the differential resistance $r_d = (dV/dI)$ of the junction. However, as far as the noise is concerned, a limitation occurs because the current generator $(4kT\Delta f I/V_0)^{1/2}$ gives rise to an equivalent e.m.f. $(4kT\Delta f I_0/V_0)^{1/2}(r^2 + \omega_s^2 L^2)^{1/2}$ in series with r and L (Fig. 16.2b). The noise figure at the frequency ω_s is therefore

$$F = 1 + \frac{4kT\Delta f(I/V_0)(r^2 + \omega_s^2 L^2)}{4kTr\Delta f} = 1 + \frac{I}{V_0}\left(r + \frac{\omega_s^2 L^2}{r}\right) \quad (16.16)$$

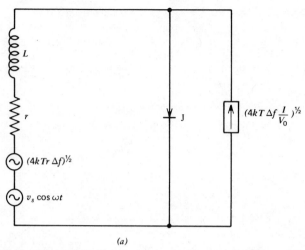

Figure 16.2. (*a*) Equivalent circuit of a Josephson junction amplifier.

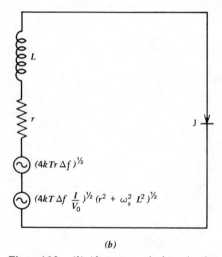

Figure 16.2. (*b*) Alternate equivalent circuit.

which has a minimum value

$$F_{\min} = 1 + 2\frac{I_0}{V_0}\omega_s L \quad \text{for} \quad r = \omega_s L \qquad (16.16a)$$

The noise figure thus increases with increasing frequency, and this sets an upper limit to the bandwidth that can be handled. Defining an upper cut-off frequency f_0 by equating

$$2\pi f_0 L = r$$

yields $f_0 = 160$ kHz for $r = 10^{-3}$ Ω and $L = 10^{-9}$ henry. The low-frequency noise figure in that case is

$$F_0 = 1 + \frac{I}{V_0}r \qquad (16.16b)$$

For a 30 MHz carrier frequency $V_0 = 0.062$ μV and I is typically of the order of 50 μA. Consequently for $r = 10^{-3}$ Ω, $F_0 = 1.8$. The noise figure F_0 decreases if the carrier frequency f_0 is increased, because that means increasing V_0.

What is the minimum power that can be handled without drowning in the noise? P_{eq} follows from

$$\overline{v_s^2} = F_0 \cdot 4kTr, \quad \text{or} \quad P_{eq} = \frac{\overline{v_s^2}}{4r} = F_0 kT \qquad (16.16c)$$

Putting $F_0 = 1.8$, $T = 4.2°$ K yields $P_{eq} = 1.06 \times 10^{-22}$ W/Hz and $(\overline{v_s^2})^{1/2} = (4rP_{eq})^{1/2} = 0.65 \times 10^{-12}$ V/Hz$^{1/2}$ for $r = 10^{-3}$ Ω.

There are various ways of improving the performance of this amplifier:

1. One can go to a higher carrier frequency f_0. This not only lowers the low-frequency noise figure F_0 but also increases the generated power, because of the increase in the d.c. voltage V_0.
2. One can decrease the inductance L of the circuit; the maximum generated power $P_{\max}$ is inversely proportional to the inductance L.

Both effects can be described by the equation

$$P_{\max} = \text{const} \frac{h}{2c} \cdot \frac{f_0}{2L} \qquad (16.17)$$

Increasing $P_{\max}$ greatly simplifies the detection of the f.m. modulated signal of carrier frequency f_0 and thus makes it easier to reach the limit set by (16.16c).

Choe, at the University of Minnesota, built a Josephson junction amplifier operating at a carrier frequency f_0 of 30 MHz.* He was unable to reach the limit set by (16.16c) because the carrier signal drowned in the noise background. In addition multiple quantum transitions occurred, and this may have considerably increased the noise. Although this noise-generation mechanism is not completely clear, the multiple quantum transition can be understood as follows:

If the junction pressure of a point-contact Josephson junction is increased, the contacting area between the two superconductors becomes wider and more critical current can flow through the junction. In such a situation multiple quantum motion $N\phi_0$, where $\phi_0 = (h/2e)$ is the flux quantum and N is an integer, is possible, As a consequence the frequency $V_0/N\phi_0$ and its harmonics are generated so that the frequency of oscillation may be written

$$f = \frac{n}{N} \frac{V_0}{\phi_0} \qquad (n=1,\ldots N) \tag{16.18}$$

The power generated for $n=1$ is much larger than the power generated for the case $V_0' = (V_0/N)$, so that it is easier to raise the signal above the noise background of the amplifier.

Operating in this manner he was able to detect a signal of about 10^{-11} V/Hz$^{1/2}$ for $r \cong 10^{-3}$ Ω.

16.2b The Josephson Junction as a Parametric Amplifier

It will now be shown that the Josephson junction exhibits a time-dependent differential inductance.

The basic Josephson junction equations are

$$I = I_0 \sin \Theta \tag{16.19}$$

$$\frac{d\Theta}{dt} = \frac{2eV_0}{\hbar} \tag{16.20}$$

where I_0 is the maximum d.c. current that can flow in the junction, V_0 is the applied voltage, $\hbar = (h/2\pi)$, and Θ is the phase difference of the wavefunction across the oxide layer between the two superconductors. Integrating (16.20) yields

$$\Theta = \frac{2eV_0}{\hbar} t + \Theta_0 \tag{16.21}$$

---

*H. M. Choe, Ph.D. Thesis, University of Minnesota, 1973.

and substituting into (16.19) gives

$$I = I_0 \sin\left(\frac{2eV_0}{\hbar} t + \Theta_0\right) \quad (16.22)$$

which corresponds to a signal of frequency $f_0 = (2eV_0/h)$. The a.c. Josephson effect is thus a direct consequence of (16.19)–(16.20).

Now the differential junction inductance L_j is defined as

$$L_j = \frac{V_0}{dI/dt} \quad (16.23)$$

Substituting (16.19) and (16.20)

$$L_j = \frac{V_0}{(I_0 \cos \Theta) d\Theta/dt} = \frac{\phi_0}{2\pi I_0 \cos \Theta} \quad (16.23a)$$

where $\phi_0 = (h/2e)$ is the elementary flux quantum. By virtue of (16.21) this changes periodically with time with an angular frequency $\omega_0 = (2eV_0/\hbar)$. This result can be applied in parametric amplification.*

16.2c The Josephson Junction as a Mixer

Because of the strongly nonlinear behavior of the Josephson junction, sum and difference frequencies will be generated if two signals of frequencies f_1 and f_2 are applied to the junction. The device can, therefore, be used as a mixer. For details see Richards et al.[†]

16.3 THE SQUID

The term "SQUID" stands for superconducting quantum interference device; it is used to measure extremely small magnetic fluxes. It is applied in a d.c. or a.c. form and is based on the fact that the flux in a superconducting ring is an integral number times $\phi_0 = (h/2e)$ (flux quantization).

16.3a The D.C. SQUID

Figure 16.3a shows the d.c. form of the SQUID; it consists of a tiny superconducting ring having an inductance L of about 10^{-9} H which

---

*P. H. Russer and H. Bayegan, *Proc. I.E.E.E.*, **61**, 46 (1973).
[†]P. L. Richards, et al., *Proc. I.E.E.E.*, **61**, 36 (1973).

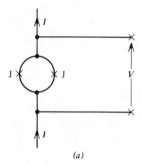

(a)   Figure 16.3. (a) Schematic of a d.c. SQUID.

contains two Josephson junctions. A d.c. current I is passed through the circuit; if I is smaller than twice the critical current I_0 of each junction, the voltage V developed across the ring is zero, whereas $V>0$ for $I>2I_0$. We assume that the circuit is operating in that condition.

If now an external flux $\phi_s < \tfrac{1}{2}\phi_0$ is passed through the ring, where $\phi_0 = (h/2e)$ is the elementary flux quantum, a screening current $I_s = (\phi_s/L)$ is set up in the ring to make the flux through the ring zero. If ϕ_s increases to $\phi_0/2$, I_s increases to $\phi_0/2L$. If the flux ϕ_s is increased further, the flux in the ring makes a transition from the $s=0$ to the $s=1$ quantum state, I_s changes to $-\phi_0/2L$ and then increases with increasing ϕ_s until ϕ_s reaches the value $3\phi_0/2$, after which point the process repeats itself. The current I_s as a function of ϕ_s thus shows the sawtooth pattern seen in Fig. 16.3b.

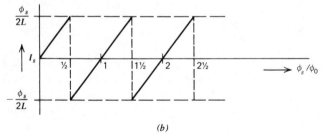

(b)

Figure 16.3. (b) I_s versus (ϕ_s/ϕ_0).

If there is a current I_s in the ring, the net current in the one junction exceeds the net current in the other-junction by an amount $2|I_s|$. The critical current of the one junction is exceeded if the other current is $I_0 - 2|I_s|$. Therefore, the total critical current is $I_0' = (2I_0 - 2|I_s|)$; I_0' is a

The Squid

maximum for $\phi_s = n\phi_0$ and a minimum for $\phi_s = (n+\frac{1}{2})\phi_0$. The I, V characteristic is thus as shown in Fig. 16.3c.

If the device is now operated at a constant current I_1 somewhat larger than $2I_0$ and ϕ_s increases gradually, the voltage V across the circuit varies periodically in a triangular fashion. The peak-to-peak amplitude is, if R is the resistance of each junction,

$$V = 2|I_s|_{\max} \frac{R}{2} = \frac{\phi_0 R}{2L} \tag{16.24}$$

since $|I_s|_{\max} = (\phi_0/2L)$. For $R = 5\,\Omega$ and $L = 10^{-9}$ henry this amounts to 5 μV since $\phi_0 = 2.07 \times 10^{-15}$ weber.

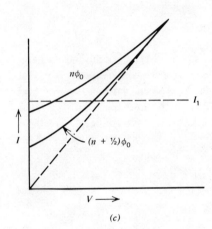

(c)

Figure 16.3. (c) Limiting I, V characteristics of the SQUID.

Therefore, a small change $\delta\phi_s$ in the flux ϕ_s produces an output voltage

$$\delta V_s = \frac{R}{L}\delta\phi_s \tag{16.25}$$

that can be detected by standard techniques. The theoretical limit to the measurement is set by noise; the practical limit is presently larger and is of the order of 10^{-3}–$10^{-4}\,\phi_0$. For details and references we refer to Clarke's review paper.*

*J. Clarke, *Proc. I.E.E.E.*, **61**, 8 (1973).

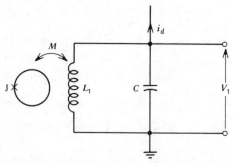

Figure 16.4. Alternating current SQUID. A coil with a single Josephson junction is coupled to an LC circuit driven by an a.c. signal i_d and the output voltage is determined as a function of the flux passing through the coil.

16.3b The R.F. SQUID

In the r.f. SQUID a tank circuit tuned at, say, 30 MHz, is coupled to a tiny loop containing a single Josephson junction. A current i_d is passed through the tank circuit (Fig. 16.4); the voltage V_T developed across the tank circuit depends on the flux ϕ_s passing through the loop.

For further details and references see Clarke's review paper.*

REFERENCE

The Jan. 1973 issue of the Proceedings of the I.E.E.E. contains many pertinent papers.

*J. Clarke, *Proc. I.E.E.E.*, **61**, 8 (1973).

17

HIGH-ENERGY QUANTUM AND CHARGED-PARTICLE DETECTORS

17.1 PRINCIPLES

When x-ray or γ-ray quanta or high-energy charged particles are absorbed in a semiconducting or insulating medium, hole–electron pairs are created. When a sufficiently large field is applied, the holes and electrons can be collected by collecting contacts and so provide pulses that can be processed by the method discussed in Chapter 8. One thus obtains *semiconductor counters*. Earlier counters used gaseous ionization chambers, but these have now been replaced by their semiconductor counterparts. The most important counters use silicon or germanium.

In an alternate arrangement no field is applied, but the material is a luminescent substance in which most of the energy is used to produce luminescence quanta that are used to excite the photocathode of a photomultiplier. Such counters are known as *scintillation counters*; the most important material is thallium-doped sodium iodide. These counters are mostly used to count γ-ray quanta.

The contacts to the semiconductor counters must be *noninjecting contacts* or *blocking contacts*. An *injecting* contact would give rise to current flow by space-charge limited flow upon application of a large d.c. voltage, and this would result in excess noise. A *blocking* contact is a contact that collects the carriers arriving at the contact but that blocks injection of carriers in the other direction. Blocking contacts can be made by using metal contacts with the proper work function, but the simplest solution is obtained by diffusing a p^+ region in on the one side and an n^+ region in on the other side and by making ohmic contacts to these regions. One then obtains a $p^+ - i - n^+$ diode that is used in the back-biased mode.

The leakage current of such devices gives shot noise; if that amount of noise becomes too large, the device must be cooled to reduce the leakage current to a sufficiently low level.

Let us assume that in either manner a back bias is applied such that the field in the i region is of the order of a few thousand V/cm and that the i region is so wide that the ionizing quanta or particles lose all their energy in the i region. If E is the energy of the particle or quantum, and E_0 the energy needed to create a hole–electron pair, the average number of pairs created is

$$\bar{n} = \frac{E}{E_0} \tag{17.1}$$

We thus see that in this case $\bar{n}$, and hence the height of the resulting pulses, depends linearly on the primary energy E.

One important characteristic of the counter is the energy E_0 needed to produce a hole-electron pair. Another important parameter of the counter is the Fano factor F, defined by the relation

$$\overline{\Delta n^2} = F\bar{n} \tag{17.2}$$

The fact that for silicon and germanium F is much smaller than unity is an indication that the creation of hole–electron pairs is not a Poisson process. Typical values for silicon and germanium are $F = 0.12$.

The corresponding uncertainty in the "measured energy" is therefore given by

$$\sqrt{\overline{\Delta E^2}} = E_0 \sqrt{\overline{\Delta n^2}} = E_0 (F\bar{n})^{1/2} \tag{17.3}$$

so that the relative uncertainty is

$$\frac{\sqrt{\overline{\Delta E^2}}}{E} = \left(\frac{FE_0}{E}\right)^{1/2} \tag{17.3a}$$

For example, if $F = 0.12$, $E_0 = 3.0$ V, and $E = 10^6$ V, the relative uncertainty is 0.6×10^{-3}. It is thus important to choose materials for which the product FE_0 is as small as possible. These considerations hold for both semiconductor counters and scintillation counters.

A third important parameter in semiconductor counters is the time τ of flight of the holes and electrons. In a p–i–n diode the field in the i region, and hence the velocity v of the particles, is constant. Consequently, if the i region has a width d, the time of flight, and hence the duration of the pulses is

$$\tau = \frac{d}{v} \tag{17.4}$$

For example, if $d = 1.0$ cm and $v = 10^6$ cm/s, $\tau = 10^{-6}$ s.

The velocity v is given by

$$v = \mu E \tag{17.5}$$

where μ is the carrier mobility and E is the field strength. For $\mu = 1000$ cm^2/V s. one needs a field strength of 10^3 V/cm to make $v = 10^6$ cm/s. Of course one can increase the field strength and so increase v, but one cannot go too far in that direction for two reasons:

1. At large field strengths the velocity v saturates to a limiting value v_{sat}, which is typically 10^7 cm/s. Raising the field strength beyond that value does not speed up the response and is actually harmful for the reason discussed under (2) below.

2. If the field strength becomes too large, collision ionization sets in, in which process the electrons and (or) holes produce additional hole–electron pairs. This is a very noisy process and should, therefore, be avoided.

In scintillation counters the important parameter is the time constant τ of the luminescence process. To obtain short light flashes one should make τ short; that is, one should use fast luminescent materials.

Semiconductor counters and scintillation counters respond to high-energy electrons and (or) high-energy quanta. They do not work for neutral particles such as neutrons, since these do not produce direct ionization. To detect neutrons, one must let them initiate nuclear reactions that result in high-energy charged particles or quanta.

For heavy charged particles, such as high-energy ions, the ions tend to become neutral toward the end of their range and cease to produce hole–electron pairs. This gives rise to an error in the measured energy that must be evaluated and corrected for.

We next discuss the saturation current I_0 of the semiconductor counter. If $G(x)$ is the generation rate of hole–electron pairs, A is the cross-sectional area of the i region and d its width, then

$$I_0 = A \int_0^d G(x) dx \tag{17.6}$$

However, according to the Shockley–Read–Hall theory

$$G(x) = \frac{n_i^2}{(n+n_1)\tau_{p0} + (p+p_1)\tau_{n0}} \quad (17.7)$$

where n_i is the intrinsic carrier concentration, n and p the electron and hole concentrations, respectively, n_1 and p_1 the carrier concentrations when the Fermi level is at the trap level, and τ_{p0} and τ_{n0} are time constants.

For intrinsic material, with the trap level at the intrinsic level $n = p = n_1 = p_1 = n_i$ and

$$G = G_0 = \frac{n_i}{2(\tau_{p0} + \tau_{n0})} \quad (17.8)$$

so that

$$I_0 = G_0 A d \quad (17.9)$$

This varies as $\exp(-E_g/2kT)$, where E_g is the energy gap of the material; I_0 can thus be considerably reduced by going to lower temperatures. This is especially important for Ge counters, since Ge has a gap width of only 0.70 V; Si has a gap width of 1.10 V, and hence I_0 is orders of magnitude lower. Nevertheless, it is also important that silicon counters are cooled in order to reduce I_0 still further.

By extreme purification one can make Ge with about 1 part in 10^{14} impurities. For Si there are technological difficulties in obtaining such low impurity and hence counter-doping techniques are applied. Commonly the lithium drift technique is applied to weakly doped p-type material. Earlier this drift technique was also applied to Ge, but in view of the very low-impurity concentrations that can now be achieved directly, this is no longer necessary.

Lithium is a donor atom that can readily diffuse into Si or Ge at elevated temperatures. For Ge the drift temperature is 20–50° C and hence one had to be careful that voltage was applied to Li-drifted Ge detectors only after cooling to a low temperature. For Si the drift temperature is 120–140° C, and hence the Li-drifted Si detectors can have voltage applied at room temperature. To drift an ideal 1 cm thick Si detector requires about 14 days at 500 V, with 1 mA current flowing through the detector.

The range of 10^8 eV α particles is about 4 mm for Si, whereas electrons have 1 cm range in Si at energies of about 4×10^6 eV.

To see what happens if the middle region is not quite intrinsic, we consider a n^+–p^- diode where the p region has an acceptor concentration $N_a = 10^{10}/\text{cm}^3$. At the metallurgical contact the space charge region then has a maximum field strength F_{max}, where

$$F_{max} = \frac{eN_a}{\varepsilon\varepsilon_0} d \qquad (17.10)$$

where d is the width of the space-charge region. Substituting $F_{max} = 10^4$ V/cm and $\varepsilon = 12$, yields $d = 6.6$ cm. Therefore, if Si detectors of 1 cm length are used, an effective acceptor concentration of $10^{10}/cm^3$ gives an almost uniform field in the i region.

17.2 APPLICATIONS

17.2a The Ge Detector

In germanium 2.96 eV energy is needed to create a hole–electron pair at 90° K. The Fano factor for large Ge detectors is 0.12, but for small Ge detectors it may be as low as 0.08; this result has not been explained. Bilger* has written a classical paper about the Fano factor in Ge. For γ rays $\sqrt{\overline{\Delta E^2}} = 1.4 \times 10^3$ eV for 10^6 eV quanta.

17.2b The Si Detector

In silicon $E_0 = 3.64$ eV at 25° C and $E_0 = 3.80$ eV at 90° K. The Fano factor for silicon is about 0.12. For 10^4 eV x rays $\sqrt{\overline{\Delta E^2}} = 160$ eV, for 50×10^6 eV α particles $\sqrt{\overline{\Delta E^2}} = 1.1 \times 10^4$ eV.

17.2c The NaI Scintillation Counter

Thallium-doped NaI scintillation detectors have a resolution of about 75×10^3 eV for 1.27×10^6 eV γ quanta, or about 40 times poorer than Li-drifted Ge detectors at those quantum energies; this must come from the product FE_0. However, since very large NaI crystals can be made, they find continued application for γ-ray quanta in excess of 10^7 eV energy. By simple optical means one can direct all the available light toward the photocathode.

One further drawback of the detector is that the uncertainty in the number of measured electrons introduced by the electronic circuitry (Chapter 8) has a larger effect than in the Si or Ge detectors, because the

---

*H. R. Bilger, *Phys. Rev.*, **163**, 238 (1967).

photocathode has a relatively low quantum efficiency η. If $\overline{\Delta N^2}$ is the variance in the measured number of electrons due to the electronic circuitry, then the resulting variance in the measured number of quanta per particle is $\overline{\Delta N^2}/\eta^2$, and hence the variance in the measured energy is $(E_0/\eta)^2 \overline{\Delta N^2}$. Typically $\eta = 0.10 - 0.20$. For a silicon or germanium detector E_0 is smaller and $\eta = 1.00$.

REFERENCES

G. Dearnaley and D. C. Northrop, *Semiconductor Counters for Nuclear Radiations*, Wiley, New York, 1966.

G. Bertolini and A. Coche (eds.), *Semiconductor Detectors*, Wiley-Interscience, New York; North Holland, Amsterdam, 1968.

Joseph Cerny (ed.), *Nuclear Spectroscopy and Reactions*, Part A (Vol. **40A** in the series *Pure and Applied Physics*), Academic, New York, 1974; see especially Sections IIIA and IIIB. Section IIID discusses the electronics used in the detection of the pulses.

APPENDIX

A.1 INTRODUCTION TO THE THEORY OF FERROELECTRICS

If we apply a field E to a linear dielectric, the dielectric becomes polarized and has a polarization P. The force acting on an ion of the lattice or on an electron bound to this ion is then not equal to eE, but must be written as eE_l, where E_l is the local field. The reason is that the other polarized ions also exert a force on the individual ions. This force is, of course, proportional to the polarization P; we may thus write this force as $(e\gamma P/\varepsilon_0)$, where γ is a proportionality factor which is $\frac{1}{3}$ for cubic symmetry but may be different for other symmetries. Hence we have

$$E_l = E + \gamma \frac{P}{\varepsilon_0} \qquad (A.1)$$

If α is the polarizability of the ions, and N the number of ions per unit volume, then the local field induces a dipole moment αE_l in each ion, and hence the polarization is

$$P = N\alpha E_l = N\alpha E + \gamma \frac{N\alpha}{\varepsilon_0} P \qquad (A.2)$$

Or, solving for P,

$$P = \frac{N\alpha/\varepsilon_0}{1 - \gamma N\alpha/\varepsilon_0} \varepsilon_0 E = (\varepsilon - 1)\varepsilon_0 E \qquad (A.3)$$

so that the susceptibility is

$$\varepsilon - 1 = \frac{N\alpha/\varepsilon_0}{1 - \gamma N\alpha/\varepsilon_0} \qquad (A.3a)$$

For $(\gamma N\alpha/\varepsilon_0) = 1$ this solution blows up; one would obtain an infinite P for finite E. Of course, this does not occur; the relationship becomes nonlinear because α decreases with increasing P. Since α cannot depend on the *sign* of P, it must be a function of P^2. We may thus write

$$\alpha = \frac{\alpha_0(T)}{1 + bP^2 + cP^4 + \ldots} \quad (A.4)$$

Substituting into (A.2) and multiplying both sides by $1 + bP^2 + cP^4 + \cdots$ yields, when collecting terms in P and in E

$$\left[1 - \frac{\gamma N\alpha_0(T)}{\varepsilon_0}\right] P + bP^3 + cP^5 + \cdots = N\alpha_0(T)E$$

or

$$E = \left[\frac{1 - \gamma N\alpha_0(T)/\varepsilon_0}{N\alpha_0(T)}\right] P + \frac{b}{N\alpha_0(T)} P^3 + \frac{c}{N\alpha_0(T)} P^5 + \ldots \quad (A.5)$$

Now $\alpha_0(T)$ is usually a monotonically decreasing function of T. There can then be a temperature T_c such that

$$\gamma \frac{N\alpha_0(T_c)}{\varepsilon_0} = 1 \quad (A.5a)$$

Let us assume that this is the case. Making a Taylor expansion around $T = T_c$ and breaking it off at the lowest nonzero terms yields

$$E = \frac{(\gamma/\varepsilon_0)(-d\alpha_0/dT)|_{T_c}}{\alpha_0(T_c)} (T - T_c) P + \frac{b}{N\alpha_0(T_c)} P^3 + \frac{c}{N\alpha_0(T_c)} P^5 + \ldots$$

$$(A.6)$$

which may be written

$$E = \beta(T - T_c)P + B_0 P^3 + C_0 P^5 \ldots \quad (A.6a)$$

with $\beta > 0$. This is the basic equation describing ferroelectric behavior and used in Chapter 12. As mentioned in that chapter, the equation $E = 0$ has a nonzero solution for $T < T_c$. In that case the material shows spontaneous polarization for $T < T_c$; that is, it is ferroelectric.

INDEX

Amplifier, 56ff, 60ff, 69ff
Amplifier noise, 56ff, 60ff, 69ff, 155ff
Autocorrelation function, 3, 7ff, 37ff
Avalanche multiplication, 136
Averages, 3ff
  ensemble, 4ff
  time, 5

Bandwidth considerations (in photoelectric detection), 129ff
Barkhausen noise, 172
Beats (between optical signals), 106, 145
Binomial distribution, 17
Binomial process, 17
Black body mode, 103
Black body radiation, 103ff
  fluctuations in, 103ff
Blocking contacts, 217
Boltzmann factor, 11
Boltzmann's constant, 10
Brownian motion, 9

Capacitance, d.c., 169
  small-signal, 160
Capacitive bolometer, 166ff
  a.c. biased, 166ff
  d.c. biased, 166ff
Carson's theorem, 42
Cathodoluminescence, 198ff
  noise, 198ff
Central limit theorem, 22
Channel multiplier, 139
Charge control devices, 182
Charged particle detectors, 217ff
  Ge, 221

  Si, 221
Chopper method, 90
Classical detectors, 140ff
Common base transistor circuit, 74
Common emitter transistor circuit, 69
Common gate FET circuit, 63
Correlation, 3
  coefficient, 6-7
  full, 6
  partial, 6
Counterdoping, 154
Counting techniques, 13ff
  current measurement by, 14ff
  radiation measurement by, 16ff
Cross-correlation function, 3, 7ff
Cross-correlation spectra, 36
Critically damped galvanometer, 13, 88ff
Curie temperature, 163
Current amplification factor, 70
Current measurements, 14ff, 88f

Dark current (in p-n diodes), 124
Darlington circuit, 75
Detectivity D^V, 100
Devonshire theory, 163, 223
Dielectric constant (small signal), 160
Distribution function, 3ff
  multivariate, 5ff
Distribution in time constants, 78
Donor levels, 153

Electron affinity, 27, 121
Electron multiplication, 25
Emitter follower, 73
Ensemble, 3

subensemble, 23
Equipartition theorem, 11, 166
Equivalent circuits, 55ff
Equivalent saturated diode current, 51
Ergodic, 5

Fano factor, 29, 218ff
Fat zero, 184ff
Fermi statistics, 18
Field effect transistor, common gate, 62
  dual gate, 66
  junction gate, 53ff
  MOS, 53ff
  tetrode, 66
Flicker noise, 2, 53, 77ff, 155ff
  in carbon resistors, 84ff
  in integrated resistors, 82
  in MOSFETs, 80ff
  in photoconductive detectors, 155ff
  in transistors, 82ff
Flux quantization, 204
Fourier analysis, 30ff
Fourier coefficient, 31ff
Fourier component, 31ff
Fourier theorem, 30
Fourier transform, 33, 42
Friiss' formula, 59

Gain, in photoconductors, 149
  in photomultipliers, 138ff
  in phototransistors, 133
Gaussian distribution, 21
Generation-recombination noise, 2, 46, 53, 85ff
  due to centers in space charge region, 87
  due to traps or impurity centers, 85
Guard ring p-n diode, 125

Heterodyne receiver, 188
Heterodyning (optical), 188ff

Image orthicon, 174ff
Image sensors, 181
  noise in, 184
Integrated resistors, 82
Intrinsic material, 152

JFET, 53ff
  generation-recombination noise in, 53, 85ff

thermal noise in, 53ff
Josephson effect, a.c., 204ff
  d.c., 204ff
Josephson junction, 204ff
  amplifier, 209ff
  mixer, 213
  parametric amplifier, 212-213
  thermometer, 205ff

Kramers-Kronig relation, 173

Langevin method, 44
Laser, 19, 105, 188
Laser radiation, 105ff
  fluctuations in, 105ff
Light amplification, 198ff
Light amplifiers, 202ff
Limiting accuracy, 14
Lithium drift technique, 220
Local field, 164, 223
Lorentz factor, 164, 223

Master equation, 48
Measurement, of current, 14ff, 88ff
  of radiation, 16ff
  of small charges, 94
  of voltages, 90
Metal-oxide-metal diode, 144
MOS capacitors, 181
MOSFETs, 53ff
  flicker noise in, 53f, 80ff
  thermal noise in, 53ff
Multiplication methods, 133ff
  avalanche, 136
  secondary emission, 138ff
Multiplication noise, 26

Networks, four-terminal, 52
  two-terminal, 50
Noise, *see* Flicker noise, Generation-recombination noise, Temperature fluctuation noise, Shot noise, and Thermal noise
Noise characterization, 50ff
Noise conductance, 50
  input, 53
Noise deterioration factor, in channel multipliers, 139
  in light amplifiers, 199
  in photomultipliers, 138
  in secondary electron conduction

Index

vidicons, 180
Noise in electrical circuits, 10ff
  in field effect transistor circuits, 60ff
  in galvanometers, 12ff
  in transistor circuits, 69ff
Noise figure, 56ff
Noise reduction methods, 157ff
Noise resistance, 50, 52
Normal distribution, 21
Normalization, 4, 7
Normal law, 21
Nyquist's theorem, 40
  quantum correction of, 41

Partially silvered mirror, 18
Partition noise, 23ff
Pentode, 17, 23
Phase sensitive detector, 91ff
Photocathode, 16
Photoconductive detectors, 146ff, 192
  noise response of, 150ff
  signal response of, 146ff
Photodiodes, 18, 119ff, 192
Photoelectric detectors, 119ff
Photoemissive cathodes, 18
Photoemissive diodes, 119ff
Photomixing, 188ff
  gain in, 189
  noise in, 190
  photoconductive, 192
  photodiode, 190
  photomultiplier, 190
  photovoltaic, 192
  Schottky barrier diode, 194
Photomultiplier, 16, 138ff, 190
Photon, 16
Phototransistor, 133
Photvoltaic cells, 119ff, 192
Planck's constant, 103, 204
p-n diode, 19, 39, 119ff
p-n junction, 39, 119ff
Poisson distribution, 19
Poisson events, 13ff
Poisson process, 19ff
Poisson's law, 19
Polarizability, 223
Polarization, 161ff
Polarization noise, 173
Power gain, 59
  available, 60

Probability density function, 3ff
  joint, 5ff
  multivariate, 5ff
Pyroelectric coefficient, 162
Pyroelectric detector, 67, 167ff
Pyroelectric vidicon, 186-187
  anode stabilization readout in, 186
  cathode stabilization readout in, 186
  chopping mode in, 186
  panning mode in, 186

Quantum detectors, 119ff, 217ff
  high energy, 217ff

Radiation resistance (of antenna), 142, 194
Random variable, 2
  continuous, 2
  discrete, 2
  stationary, 2
Resistive bolometer, 111ff
Richardson's law, 16
Rise time (in photodiodes), 128

Schottky barrier diode, 141ff, 194ff
  detector, 141ff
  mixer, 194ff
Schottky's theorem, 39
Scintellation counter, 217
  NaI, 221
Secondary electron conduction vidicon, 179ff
Secondary electron multiplication, 14, 28, 138ff
Secondary emission multiplier, 14ff
Secondary emission noise, 28
Semiconductor counters, 218ff
Shockley-Read-Hall centers, 152
Shockley-Read-Hall recombination, 220
Shot noise, 1, 39, 119ff, 136, 139
  in diodes, 121ff
  in transistors, 39
Source follower, 62
Spectral intensity, 31f, 34ff
  cross-, 36
  self, 36
SQUIDD, d.c., 213ff
  R.F., 216
Stationary processes, 4, 6
Stephan-Boltzmann constant, 9, 102
Stephan-Boltzmann's law, 102

Superconductivity, 204
Superpoissonian, 19, 28

Technical sensitivity, 12, 99ff
Television pick-up tubes, 174ff
Temperature fluctuation noise, 2, 45, 55, 101ff
Thermal noise, 10, 40, 44, 53
 in FETs, 53
 in photoconductors, 155
Thermal radiation detectors, 99ff
Thermionic diode, 19, 39
 saturated, 39
 space charge limited, 39
Thermocouple detector, 106ff
Transconductance, 53
Transfer admittance, 52
Transistors, 18, 23, 39, 69ff
 flicker noise in, 82ff
 shot noise in, 39
Trapping, 78
Tunneling, 78
 probability, 78
Two-stage amplifier, 65, 72

Variance, 5
Variance theorem, 22ff
Vidicon, 177ff
 pyroelectric, 186ff
 secondary electron conduction, 179ff
Voltage measurements, 90ff

Wiener Khintchine theorem, 31ff

Zero point energy, 103